Lecture Notes in Applied and Computational Mechanics

Volume 46

Series Editors

Prof. Dr.-Ing. Friedrich Pfeiffer
Prof. Dr.-Ing. Peter Wriggers

Lecture Notes in Applied and Computational Mechanics

Edited by F. Pfeiffer and P. Wriggers

Mechanics of Microstructured Solids

Cellular Materials, Fibre Reinforced Solids and Soft Tissues

J.-F. Ganghoffer and Franco Pastrone (Eds.)

 Springer

Professor Jean-François Ganghoffer
LEMTA - ENSEM, France
2, Avenue de la Forêt de Haye
BP 160 - 54504 Vandoeuvre Cedex
France
Email: Jean-francois.Ganghoffer@ensem.inpl-nancy.fr

Professor Franco Pastrone
Department of Mathematics
University of Torino
Via Carlo Alberto, 10
10123 Torino
Italy
Email: franco.pastrone@unito.it

ISBN: 978-3-642-10161-8   e-ISBN: 978-3-642-00911-2

DOI 10.1007/978-3-642-00911-2

Lecture Notes in Applied and Computational Mechanics   ISSN 1613-7736
                                                         e-ISSN 1860-0816

Typeset & Cover Design: Scientific Publishing Services Pvt. Ltd., Chennai, India.

Printed on acid-free paper

9 8 7 6 5 4 3 2 1 0

springer.com

Preface

This special volume of the series *Lecture Notes in Applied and Computational Mechanics* is a compendium of reviewed articles presented at the 11th EUROMECH-MECAMAT conference entitled "Mechanics of microstructured solids: cellular materials, fibre reinforced solids and soft tissues", which took place in Torino (Italy) in March 10-14, 2008, at the Museo Regional delle Scienze. This EUROMECH-MECAMAT conference was jointly organized by the Dipartimento di Matematica dell'Università di Torino, Italy and the INPL Institute (LEMTA, Nancy-Université, France). Prof. Franco Pastrone and Prof. Jean-François Ganghoffer were the co-chairmen.

The conference brought together 50 scientists from 11 European countries, and was aimed at defining the current state of the art in the growing field of cellular and fibrous materials in Europe. Participants had interests in the constitutive models of micro-structured solids, non-linear wave propagation, setting up of models and identification of fibre reinforced solids, and soft tissue behaviour in a biomechanical context.

The conference covered most of the mechanical and material aspects, grouped in the following four sessions:

- Fibre reinforced materials;
- Soft biological tissues;
- Generalized continua: models and materials;
- Non-linear wave propagation.

The high quality talks showed a good balance between modelling and material aspects. An important part of the colloquium, with 12 presentations, was devoted to various aspects of the biomechanics of soft tissues, such as cell adhesion, constitutive models of soft tissues (brain; arteries), or models of blood flow.

According to the Euromech rules, two awards have been assigned for the best oral presentation and the best poster presentation to young researchers. The prizes were awarded to Polina Dyatlova, from Saint Petersburg, Russia, and Merle Randruup, From Tallinn, Estonia, with the following motivations:

1. **P. Dyatlova:** The presentation (Non linear problems of fiber reinforced soft shells) concerned the study of the interaction of soft shells with fibers or membranes covering the bodies. The case of a fiber net with higher rigidity

than the shell underlying is studied also numerically, by means of finite elements method, and the convergence of the procedures is discussed, very clearly as well as the difficulties of the mathematical problems that arise and the possible further developments.

2. M. Randruup: The presentation (On modelling deformation waves in microstructured materials: evolution equations) concerned the study of longitudinal deformation waves in microstructured solids., in nonlinear case. Hierarchical governing equations are used to obtain evolution equations. The results have been exploited also numerically and simulations has been showed.

Beyond the scientific reports and the poster presentations, many informal exchanges were possible mainly during the coffee breaks and the lunches, according to the scope and the spirit of Euromech Colloquia and Conferences. Surely the meeting stimulated current researches and often the discussions were lively, informal and penetrating.

A social dinner was also offered to the participants, after a visit to the Reggia di Venaria Reale, noticeable Royal Palace recently restored and brought back to the original splendour, Unesco World Heritage.

Contents

Class-Jump Phenomenon for Physical Symmetries in Bi-dimensional Space

N. Auffray, R. Bouchet, and Y. Bréchet

In condensed matter physics, tensors are used to model physical properties of materials. Some well known examples are elasticity which is modeled by a $4th$-order tensor, or piezoelectricity by a $3rd$-order tensor. Some tensors of higher orders might occur in theory of generalized continuum, for example, in Mindlin's strain-gradient elasticity [2], physical state of the material is described in each point using three tensors from $4th$-order to $6th$-order. Damage could also be modeled by an $8th$-order tensor. When a complex behavior is studied the first step is to focus attention on its bi-dimensional version. Taking a complex problem at a lower level helps to better understand the physics of the behavior. But it has to be wondered in which measure the phenomenon studied is independent of the dimension of the working space. In this paper attention will be focused on the issue of symmetry classes of linear constitutive behavior, and it will be shown that, in $2D$, anomalies could occur.

1 Physical and Material Symmetries

In the following, $\mathcal{E}^d$ will be the d-dimensional euclidean space. The physical space will be modeled by $\mathcal{E}^d$. The cases $d = 2$ and $d = 3$ will then be considered.

Let G be a group of operations, a material $\mathcal{M}$ is said to be G-invariant if the action of the elements of G transforms $\mathcal{M}$ into itself. This set of operations, namely the material symmetry group, will be noted $G_{\mathcal{M}}$.

N. Auffray and R. Bouchet
ONERA/DMSM, 29 Avenue de la Division Leclerc,
F-92322, Châtillon Cedex, France
e-mail: `nicolas.auffray@onera.fr`

Y. Bréchet
LTPCM, BP 75, Domaine Universitaire de Grenoble, F-38402,
Saint Martin d'Hères Cedex, France

J.-F. Ganghoffer, F. Pastrone (Eds.): Mech. of Microstru. Solids, LNACM 46, pp. 1–11.
springerlink.com   © Springer-Verlag Berlin Heidelberg 2009

$$G_{\mathcal{M}} = \{Q \in O(d), \quad Q \star \mathcal{M} = \mathcal{M}\} \tag{1}$$

where $\star$ represents the action of Q upon $\mathcal{M}$ and $O(d)$ is the orthogonal group in d-dimension.

Consider now a physical property $\mathcal{P}$ defined on a material. The physical symmetry group is defined as the set of operations that leave the behavior invariant. This group will be noted $G_{\mathcal{P}}$.

$$G_{\mathcal{P}} = \{Q \in O(d), \quad Q \star \mathcal{P} = \mathcal{P}\} \tag{2}$$

This physical property would be described, in the following, by an nth-order tensor noted $\mathrm{T}^{(n)}$. $\mathbb{T}^{(n)}$ will stand for its related vector space. The action of $O(d)$ on $\mathrm{T}^{(n)}$ is defined by the mean of the Rayleigh product:

$$\star : O(d) \times \mathbb{T}^{(n)} \to \mathbb{T}^{(n)} : (Q, \mathrm{T}^{(n)}) \mapsto Q \star \mathrm{T}^{(n)} := Q_{i_1 j_1} \cdots Q_{i_n j_n} T^{(n)}_{j_1 \dots j_n} \tag{3}$$

So the symmetry group of $\mathrm{T}^{(n)}$ is the following set of elements:

$$G_{\mathrm{T}^{(n)}} = \{Q \in O(d) | Q_{i_1 j_1} \cdots Q_{i_n j_n} T^{(n)}_{j_1 \dots j_n} = T^{(n)}_{i_1 \dots i_n}\} \tag{4}$$

According to the different symmetry properties of its elements, the space $\mathbb{T}^{(n)}$ is divided in equivalence classes [3].

The material symmetry group and the physical one are related by the mean of Neumann's principle [4]. In this way we get the inclusion:

$$G_{\mathcal{M}} \subseteq G_{\mathcal{P}} \tag{5}$$

This relation means that every operations that leave the material invariant let the physical properties invariant. Nevertheless, as shown for tensorial properties by Hermann's theorems [6], physical properties could be more symmetrical than the material.

In $\mathcal{E}^2$, plane and spatial material symmetry groups coincide meanwhile they are distinct in $\mathcal{E}^3$. In case of plane invariance, $G_{\mathcal{M}}$ should be conjugate with a subgroup of $O(2)$ [4]. The collection of those subgroups is [3]

$$\Sigma := \{I, Z_p, D_p, SO(2), O(2)\} \tag{6}$$

where I is the identity group; Z_p is the cyclic group of order p, i.e. the symmetry group of a p-fold-invariant chiral figure; D_p is the dihedral group of order $2p$, i.e. the symmetry group of a regular p-gone[1]; $SO(2)$ is the continuous group of rotations and $O(2)$ is the continuous group of orthogonal transformations in two dimension. In the sequel we will often refer to rotational invariance, according to former definitions, it will mean the collection of $SO(2)$-subgroups.

---

[1] D_p contains Z_p and mirror symmetry.

To study the symmetry classes of tensors a decomposition in elementary parts is needed. Such a decomposition is referred in literature as the harmonic decomposition [3, 1] or the irreducible decomposition [8, 7].

2 Harmonic Decomposition

2.1 Generalities

We call harmonic decomposition the orthogonal irreducible decomposition of a tensor. In $\mathcal{E}^d$ this decomposition is $O(d)$-invariant. This decomposition is well known in group representation theory and allows us to decompose a tensor of any finite order as a sum of irreducible ones [7, 8]. This decomposition could be written:

$$\mathrm{T}^{(n)} = \sum_{k,\tau} \mathrm{D}(n)^{k,\tau} \tag{7}$$

where tensors $\mathrm{D}(n)^{k,\tau}$ are components of the irreducible decomposition, k denotes the order of the harmonic tensor embedded in $\mathrm{D}(n)$ and τ separates terms of same order. This decomposition establish an isomorphism between $\mathbb{T}^{(n)}$ and a direct sum of harmonic tensor vector spaces $\mathbb{H}^k$ [3]. We shall write

$$\mathbb{T}^{(n)} = \bigoplus_{k,\tau} \mathbb{H}^{k,\tau} \tag{8}$$

but, as explained in [1], this decomposition is not unique. But the following decomposition, which groups terms of same order, is unique and is called the $O(d)$-isotypic decomposition:

$$\mathbb{T}^{(n)} = \bigoplus_{k=0}^{n} \alpha_k \mathbb{H}^k \tag{9}$$

where α_k is the multiplicity of $\mathbb{H}^k$ in the decomposition. Harmonic tensors are totally symmetric and traceless, and the dimension of the associated vector spaces are:

$$\dim \mathbb{H}^k = \begin{cases} 2k+1, & \forall\ k \geq 0 \text{ in } 3D \\ 2, & \forall\ k > 0, \text{ and } 1 \text{ if } k = 0 \text{ in } 2D \end{cases} \tag{10}$$

For the sake of simplicity when there is no risk of misunderstanding, spaces $\alpha_k \mathbb{H}^k$ would be noted K^{α_k} : i.e. the order K of the subspace powers its multiplicity α_k. Moreover, as we are dealing with $2D$ and $3D$ tensor spaces, in the notation the two kinds of vector spaces will be distinguished by an $*$ exponent for the bi-dimensional one. The bi-dimensional harmonic decomposition of $\mathbb{T}^{*(n)}$ will be written:

$$\mathbb{T}^{*(n)} = \bigoplus_{k=0}^{n} \alpha_k^* \mathbb{H}^{*k} \tag{11}$$

Families $\{\alpha_k\}$ and $\{\alpha_k^*\}$ are functions of the order of the tensor space and of its index symmetries. There exists several methods [8, 7, 5] to compute those different families. A general result on the structure of $2D$ and $3D$ harmonic decomposition can be found in [7]. For a generic tensor, i.e. a tensor space with no index symmetry, the two following theorems hold true:

Theorem 1. *Let $\mathbb{T}^{*(n)}$ be a $2D$ generic nth-order tensor vector space. If $n = 2q$ its harmonic decomposition only contains subspaces of even-order, and if $n = 2q + 1$ its harmonic decomposition only contains subspaces of odd-order.*

Theorem 2. *Let $\mathbb{T}^{(n)}$ be a $3D$ generic nth-order tensor. Its harmonic decomposition contains subspaces of kth-order for any k with $0 \leq k \leq n$.*

Taking into account index symmetries can make odd or even components to vanish in the $3D$ case. For example, a totally symmetric tensor of even-order would just contain harmonic tensors of even-order and reciprocally for totally symmetric odd-order tensor [8].

A classical example is the space of elasticity tensors in $3D$ [3]: this space is isomorphic to $0^2 \oplus 2^2 \oplus 4$ [3] which is a 21-dimensional vector space. In $2D$, the same space will be isomorphic to $0^{*2} \oplus 2^* \oplus 4^*$ [7]. This space is a 6-dimensional vector space. The information provided by these two decompositions does not allow a direct comparison of theirs structures. In order to do that, the O(2)-isotypic decomposition of $\mathbb{T}^{(n)}$ should be expressed.

We will now focus on the link between $3D$ harmonic decomposition and $2D$ one. Our aim is, for rotational action, to establish an isomorphism between $3D$ harmonic tensor vector space and a direct sum of $2D$ harmonic tensor vector spaces.

2.2 Cartan Decomposition

As shown and demonstrated in [1] under the action of O(2) the space $\mathbb{H}^k$ admits the following decomposition:

$$\mathbb{H}^k = \bigoplus_{j=0}^{k} \mathbb{K}_j^k \tag{12}$$

where

$$\dim \mathbb{K}_j^k = \begin{cases} 1 \text{ if } j = 0 \\ 2 \text{ if } j \neq 0 \end{cases} \tag{13}$$

This decomposition is referred, in the literature, as the Cartan decomposition of an harmonic tensor space. The relation (12) implies a decomposition containing subspaces for each j between 0 and k.

For $(x, y, z) \in \mathbb{R}^3$, let $w = x + iy$ be the plane complex vector with $i = \sqrt{-1}$. As explained in [1] spaces $\mathbb{K}_j^k$ are spanned by

$$k_l^j = \{k_1^j, \quad k_2^j\} = \{z^{k-j}\Re(w^j), \ z^{k-j}\Im(w^j)\} \tag{14}$$

where $\Re, \Im$ are functions returning the real and imaginary part of a complex number. In the same time spaces $\mathbb{H}^{*j}$ are spanned [7] by

$$h_l^j = \{h_1^j, \quad h_2^j\} = \{\Re(w^j), \ \Im(w^j)\} \tag{15}$$

So, if $z \neq 0$ then for each space $\mathbb{K}_j^k$ there exists a bijective function ϕ_j^k that turns $\mathbb{K}_j^k$ basis vectors into $\mathbb{H}^{*j}$ ones. ϕ_j^k is defined as

$$\phi_j^k(k_l^j) = \frac{k_l^j}{z^{k-j}} = h_l^j \tag{16}$$

By the family $\{\phi_j^k\}_{0 \leq j \leq k}$ the space $\mathbb{H}^k$ is isomorphic to the following space:

$$\mathbb{H}^k = \bigoplus_{j=0}^{k} \mathbb{H}^{*j} \tag{17}$$

The relation (9) may be rewritten

$$\mathbb{T}^{(n)} = \bigoplus_{k=0}^{n} \alpha_k \left(\bigoplus_{j=0}^{k} \mathbb{H}^{*j} \right) = \bigoplus_{k=0}^{n} \sum_{j=k}^{n} \alpha_j \mathbb{H}^{*k} = \bigoplus_{k=0}^{n} \beta_k \mathbb{H}^{*k} \tag{18}$$

This decomposition is the $O(2)$-isotypic decomposition of a three dimensional tensor. We obtain the following lemma for the planar decomposition of a $3D$ operator:

Lemma 1. *Let $\mathbb{T}^{(n)}$ be a 3D nth-order tensor vector space with any kind of index symmetry. Its planar harmonic decomposition contains subspaces of kth-order for any k within $0 \leq k \leq n$.*

Proof. The planar decomposition can be written:

$$\mathbb{T}^{(n)} = \bigoplus_{k=0}^{n} \beta_k \mathbb{H}^{*k} \tag{19}$$

So, to demonstrate this lemma, the following property shall be proved:

$$\forall k \in [0, n], \ \beta_k > 0 \tag{20}$$

The proof is straight forward. Let's consider the harmonic decomposition of a tensor space $\mathbb{T}^{(n)}$. Regardless of the index symmetry of that space, we got a term of order n in this decomposition. By property of harmonic decomposition, the multiplicity of this harmonic space must be 1. By the mean of Cartan decomposition, this space will be decomposed into $n + 1$ Cartan subspaces. We get:

$$\forall k \in [0, n], \ \beta_k = \sum_{i=k}^{n} \alpha_i = 1 + \sum_{i=k}^{n-1} \alpha_i \tag{21}$$

and as $\alpha_i \in \mathbb{N}$, we can conclude that:

$$\forall k \in [0, n], \ \beta_k \geq 1 \tag{22}$$

which concludes the proof. $\qquad\qquad\qquad\qquad\qquad\qquad\qquad\qquad\qquad\qquad\quad\square$

So theorem 1 and lemma 1 allow us to formulate the following theorem:

Theorem 3. *Let $\mathbb{T}^{(n)}$ be any nth-order tensor vector space. In $\mathcal{E}^3$ its planar harmonic spectrum is full, meanwhile in $\mathcal{E}^2$ its planar harmonic spectrum is sparse according to the parity of n.*

If we come back to the example of elasticity, we remind that, in $3D$, its vector space is isomorphic to $0^2 \oplus 2^2 \oplus 4$. According to the formula given in (18), its planar decomposition would be $0^{*5} \oplus 1^{*3} \oplus 2^{*3} \oplus 3^* \oplus 4^*$. This decomposition is the one to compare with the bi-dimensional harmonic decomposition which is $0^{*2} \oplus 2^* \oplus 4^*$.

This difference in the composition of the planar spectrum will lead to different systems of symmetry in $2D$ and $3D$ physical spaces. This can be shown by the study of rotational invariance properties of tensors.

3   Rotational Invariance

Let $\mathrm{G}_{\mathrm{T}^n}$ be the group of operations that leave T^n unchanged, i.e.:

$$Q \in \mathrm{G}_{\mathrm{T}^n} \Rightarrow Q \star \mathrm{T}^n = \mathrm{T}^n \tag{23}$$

The space $\mathbb{T}^{(n)}$ is isomorphic with $\bigoplus_{k=0}^{n} \beta_k \mathbb{H}^{*k}$. So, any $\mathrm{T}^n \in \mathbb{T}^{(n)}$ is defined by a family of tensors $\{\mathrm{H}^{*k,\tau}\}$. The order of this family is obviously $m = \sum_{k=0}^{n} \beta_k$. As this decomposition is $SO(2)$-invariant, the invariance condition on T^n could be expressed by m conditions on the elements of the planar decomposition. These m conditions are of $n + 1$ different types according to the order of the bi-dimensional harmonic tensor, i.e.

$$Q \star_k \mathrm{H}^{*k} = \mathrm{H}^{*k} \tag{24}$$

where $\star_k$ is the action of $SO(2)$ on $\mathbb{H}^{*k}$, this action shall easily be expressed in the sequel.

Let $\mathrm{H}^{*k} = (s_k, t_k)$ be a vector of $\mathbb{H}^{*k}$. Consider a plane rotation $Q_{rot} \in SO(2)$, for example, take the generator of Z_p, we get:

$$Q_{rot} = \begin{pmatrix} \cos\frac{2\pi}{p} & -\sin\frac{2\pi}{p} \\ \sin\frac{2\pi}{p} & \cos\frac{2\pi}{p} \end{pmatrix} \tag{25}$$

As shown in [1], Q_{rot} acts on $\mathbb{H}^{*k}$ as a generator of $Z_{\frac{p}{k}}$. The order of the rotation p is divided by the indice of Cartan subspace.

$$Q_{rot} \star H^{*k} = \begin{pmatrix} \cos\frac{2k\pi}{p} & -\sin\frac{2k\pi}{p} \\ \sin\frac{2k\pi}{p} & \cos\frac{2k\pi}{p} \end{pmatrix} \begin{pmatrix} s_k \\ t_k \end{pmatrix} \tag{26}$$

So if Q_{rot} belongs to G_{T^n} then each $\mathbb{H}^{*k}$ must be Q_{rot}-invariant. The matrix of the action of Q_{rot} on $\mathbb{H}^{*k}$ will be noted $Q_{rot}^{(k)}$. The invariance condition of $\mathbb{H}^{*k}$ is the solution of $(Q_{rot}^{(k)} - \text{Id})\mathbb{H}^{*k} = 0$. In other words $\ker(Q_{rot}^{(k)} - \text{Id})$ has to be studied. A direct calculation shows that the invariance condition of $\mathbb{H}^{*k}$ under the Z_p-action is:

$$k = tp, \quad t \in \mathbb{N} \tag{27}$$

And so if $k \neq tp$ then $\mathbb{H}^{*k}$ equals 0. So the planar decomposition of a Z_p-invariant space $\mathbb{T}^{(n)}$, $^{Z_p}\mathbb{T}^{(n)}$, will be:

$$^{Z_p}\mathbb{T}^{(n)} = \bigoplus_{0 \leq m \leq \lfloor \frac{n}{p} \rfloor} \beta_{mp}\mathbb{H}^{*mp} \tag{28}$$

where $\lfloor . \rfloor$ is the floor function. It is clear that in the case of $p > n$ the decomposition (28) will reduced to

$$^{Z_p}\mathbb{T}^{(n)} = \beta_0 \mathbb{H}^{*0} \tag{29}$$

which only contains the hemitropic components. The relation (28) allows us to compare consequences of material invariance on operators in different cases.

For tensors in $3D$, we got, as a consequence of theorem 1, the following theorem:

Theorem 4. *Let's consider the vector space $\mathbb{T}^{(n)}$ in $3D$. Non-equivalent rotational material invariances of non null order lower than or equal to n imply non-equivalent tensorial invariances.*

Proof. As we are dealing with a tensorial vector space in $3D$ we know that its planar decomposition is complete. For any p and q greater than n, we got $^{Z_p}\mathbb{T}^{(n)} = ^{Z_q}\mathbb{T}^{(n)} = \beta_0\mathbb{H}^{*0}$. And so we conclude that this condition leads to the same anisotropic class: transverse hemitropy[2]. This property is related to Hermann's theorems [5]. Now, suppose that p and q are both less than or equal to n. Let's search $p, q \in \mathbb{N}^{*2}$ verifying $^{Z_p}\mathbb{T}^{(n)} = ^{Z_q}\mathbb{T}^{(n)}$. The following relation could be written:

$$\bigoplus_{0 \leq m \leq \lfloor \frac{n}{p} \rfloor} \beta_{mp}\mathbb{H}^{*mp} = \bigoplus_{0 \leq m' \leq \lfloor \frac{n}{q} \rfloor} \beta_{m'q}\mathbb{H}^{*m'q} \tag{30}$$

---

[2] The mirror invariance should be combined with the rotational one to obtain the transverse isotropy class.

this relation implies that the following indices equality holds: $mp = m'q$, $\{m, m', p, q\} \in \mathbb{N}^2 \times \mathbb{N}^{*2}$.

$$\forall m, \ \exists m', \ m' = m\frac{p}{q} \tag{31}$$

and so $\frac{p}{q} = k, k \in \mathbb{N}^*$. Reciprocally

$$\forall m', \ \exists m, \ m = m'\frac{q}{p} \tag{32}$$

and so $\frac{q}{p} = k' = \frac{1}{k}, k' \in \mathbb{N}^*$, which implies that k must be 1. The only solution is $p = q$, so reciprocally $p \neq q$ implies $^{Z_p}\mathbb{T}^{(n)} \neq^{Z_q} \mathbb{T}^{(n)}$. $\square$

As an example of what this property implies, let's consider a $6th$-order tensor. For a physical illustration, this tensor could be the second order elasticity tensor that appears in strain-gradient elasticity [2]. By the mean of Hermann's theorems [6], we know that for any material invariance of order greater that 6 this tensor will be transverse hemitropic. By the mean of theorem 4, we know that any other rotational invariance leads to distinct classes of symmetry. And so, if we just consider subgroups of SO(2), we got 7 different rotational invariant types of tensors: $\{I, Z_2, Z_3, Z_4, Z_5, Z_6, SO(2)\}$. Let's consider now the same problem in a two dimensional physical space.

4   Bi-dimensional Physical Space

To study the plane invariance in $3D$ we completed the use of the harmonic decomposition by the Cartan one. In $2D$, the problem is simpler and the vector space $\mathbb{T}^{*(n)}$ should not be decomposed any further. The action on Q_{rot} on $2D$ harmonic decomposition is the same as the one introduced for the Cartan decomposition in the $3D$. And so, all the things we said about the condition of invariance remain the same. The only difference arises from theorem 1: in $2D$ the harmonic decomposition of an even-order tensor just contains even-order harmonic tensors and reciprocally. This property will generate a phenomenon, we call "Class-Jump", that makes some physical properties more symmetrical that they should.

4.1   Even-Order Tensors

Consider first the case of the vector space $\mathbb{T}^{*(2q)}$, according to theorem 1 its O(2)-isotypic decomposition could be written:

$$\mathbb{T}^{*(2q)} = \bigoplus_{k=0}^{q} \alpha^*_{2k}\mathbb{H}^{*2k} \tag{33}$$

Consider the subset of Z_p-invariant tensors. For these tensors the non-vanishing coefficients would belong to subspaces of order $2k$ verifying the relation:

$$2k = tp, \quad t \in \mathbb{N} \tag{34}$$

Two cases should be considered: $p = 2q'$ and $p = 2q' + 1$. Suppose first $p = 2q'$, the classical condition $k = tq'$ is obtained. But considering the case $p = 2q' + 1$ we obtain:

$$2k = t(2q' + 1), \quad t \in \mathbb{N} \tag{35}$$

The former relation makes sense only for $t = 2t'$, and so:

$$2k = t'(2(2q' + 1)), \quad t' \in \mathbb{N} \tag{36}$$

which is the same restriction as the one imposed by an $Z_{(2(2q'+1))}$-material invariance. A $Z_{(2q'+1)}$-material invariance will impose the same restriction on tensors as a $Z_{(2(2q'+1))}$-invariance. This demonstrates the following theorem:

Theorem 5. *In 2D, an odd-order material invariance could not be seen by an even-order. The material-invariance seen would be twice the order of the real one.*

Now consider the case of odd-order tensors.

4.2 Odd-Order Tensors

According to theorem 1, the O(2)-isotypic decomposition of the vector space $\mathbb{T}^{*(2q+1)}$ could be written:

$$\mathbb{T}^{*(2q+1)} = \bigoplus_{k=0}^{q} \alpha^{*}_{2k+1} \mathbb{H}^{*2k+1} \tag{37}$$

Consider the subset of Z_p-invariant tensors. For these tensors the non-vanishing coefficients would belong to subspaces of order $2k + 1$ verifying the relation:

$$(2k + 1) = tp, \quad t \in \mathbb{N} \tag{38}$$

Consider first the case $p = 2q'$, we obtain

$$(2k + 1) = 2tq', \quad t \in \mathbb{N} \tag{39}$$

As $t \in \mathbb{N}$, the former relation could never be satisfied. So the following theorem has been demonstrated:

Theorem 6. *In 2D, an even-order material invariance could not be seen by an odd-order tensor. The tensor would vanish for that kind of material invariance.*

Consider now $p = 2q' + 1$. We get the relation

$$(2k + 1) = t(2q' + 1), \quad t \in \mathbb{N} \tag{40}$$

If t is supposed to be even, then the relation (39) is found. It could be concluded that there is no solution for $t = 2t'$. Suppose now $t = 2t' + 1$, we obtain:

$$(2k + 1) = (2t + 1)(2q' + 1) \tag{41}$$

which is a relation of the form $k' = t'r'$ for $\{k', t', r'\} \in (2\mathbb{N} + 1)$. In this case we don't have any specific behavior for the symmetry class. Distinct material invariances will lead to distinct tensorial classes.

4.3 Synthesis

The results obtained for rotational invariance in $2D$ could be summed up in the following array:

	$\mathbb{T}^{*(2k)}$	$\mathbb{T}^{*(2k+1)}$
Z_{2p}	$t \in \mathbb{N}$	$t = \emptyset$
Z_{2p+1}	$t \in (2\mathbb{N}) \Rightarrow$ jumps to $Z_{2(2p+1)}$	$t \in (2\mathbb{N} + 1)$

The point to stress here is the class jump phenomenon that would occur for an odd material invariance in the case of a even tensor, and the vanishing effect in the opposite case. As demonstrated by theorem 4, none of these effects could appeared in $3D$.

Let's come back to the $6th$-order tensor example. In §3 it was shown that in $3D$ there exists 7 different rotational invariant types of tensors. By the mean of theorem 5, we claim that such a tensor in $2D$ could only belongs to one of the following 4 different systems: $\{Z_2, Z_4, Z_6, \mathrm{SO}(2)\}$. In particular it should be noted that, in $2D$, a tensorial property defined on a material with no symmetry is Z_2-invariant.

5 Conclusion

It has been shown in this paper that the symmetry class of a tensorial property depends on the space dimension. In $2D$, some jumps of symmetry could occur making physical properties more symmetrical than the material they are defined on. This point is related to the reciprocity of Hermann's theorems [6]. Those theorems state that if the order of the rotational invariance exceeds the number of indexes of a tensor then this tensor is, a least, transverse hemitropic (SO(2)-invariant). For the $3D$ case, theorems 1 and 4 show that the reciprocity is true. But as corollary of theorem 5 the reciprocity fails in $2D$. If we consider a $6th$-order tensor then a Z_5-material invariance will be seen as a Z_{10}-one by the tensor. And so, a Z_5-material invariance implies a $6th$-order tensor to be at least hemitropic. Continuous classes of symmetry could appear for an order of rotation lower than the number of index of the tensor.

References

1. Golubitsky, M., Stewart, I., Schaeffer, D.: Singularities and Groups in Bifurcation Theory, vol. II. Springer, Heidelberg (1989)
2. Mindlin, R., Eshel, N.: Int. J. Solids Struct. 4, 109–124 (1968)
3. Forte, S., Vianello, M.: J. Elastic. 43, 81–108 (1996)
4. Zheng, Q., Boehler, J.: Acta Mecha. 102, 73–89 (1994)
5. Auffray, N.: CR. Meca. 336, 458–463 (2008)
6. Auffray, N.: CR. Meca. 336, 370–375 (2008)
7. Zou, W., Zheng, Q., Du, D., Rychlewski, J.: Math. Mech. Solids 6, 249–267 (2001)
8. Jerphagnon, J., Chemla, D., Bonneville, R.: Adv. Physics. 27, 609–650 (1978)

Analysis and Optimisation of Interfaces for Multi-material Structures

Franz-Joseph Barthold and Monika Rotthaus

Abstract. This paper is concerned with aspects of structural analysis for interfaces using non-conforming mesh methods. Additionally, the corresponding sensitivity analysis and the optimisation procedure are outlined. Generally, strains and stresses at the material interfaces must be correctly computed and efficiently controlled so that stress concentrations yielding damage and failure are avoided. Discontinuous physical behaviour such as cracks and inclusions can be modelled on a fixed grid using the extended finite element method (XFEM). A novel Nitsche-type version (XFEM-Nitsche) capturing weak discontinuities at interfaces is presented.

1 Introduction

Non-conforming strategies, which offer more flexibility compared to traditional CAGD-based techniques, are a challenging approach in structural optimisation for interface problems. A combined treatment of topology and shape optimisation is desired. Thus, the optimal placement of additional inclusions into the matrix material, the deletion of redundant inclusions and the best geometrical shape of the interfaces are sought-after. Furthermore, optimal inclusions enforce stress control on the interface in order to avoid damage and delamination. Thus, the accurate computation of the physical quantities on the interface is a prerequisite for successful optimisation.

The standard finite element method is based on *computer aided geometric design* (CAGD), i.e. the domains are modelled by geometric curves, surfaces and volumes,

Franz-Joseph Barthold
Numerical Methods and Information Processing, Technische Universität Dortmund
August-Schmidt-Str. 8, D-44221 Dortmund, Germany
e-mail: `franz-joseph.barthold@tu-dortmund.de`

Monika Rotthaus
Numerical Methods and Information Processing, Technische Universität Dortmund
August-Schmidt-Str. 8, D-44221 Dortmund, Germany
e-mail: `monika.rotthaus@tu-dortmund.de`

J.-F. Ganghoffer, F. Pastrone (Eds.): Mech. of Microstru. Solids, LNACM 46, pp. 13–20.
springerlink.com © Springer-Verlag Berlin Heidelberg 2009

and on the associated conforming mesh. Thus, the interfaces coincide with element edges and the choice of equal shape functions yield a continuous displacement field.

Domain decomposition techniques relax the strong coupling conditions along the interfaces to obtain more flexible approaches, see e.g. [9] for an overview. The *mortar method*, see [6], offers a general tool to deal with the coupling conditions as an additional constraint applied to a non-conforming mesh. A version based on *Lagrange multipliers* leads to a saddle point problem, see [3]. The deficiencies can be overcome by a combination of the mortar method with Nitsche's method [11], see e.g. [14] for computational issues. This approach has been successfully applied to contact mechanics, see e.g. [15] for detailed hints.

An alternative approach has been formulated as *extended finite element method* (XFEM) in [4] for crack problems. The central idea is to model the geometrical phenomena of crack opening by a discontinuous displacement field. Special finite elements have been developed, which allow the discontinuity line to run across the element domain. Thus, it is desirable to use a finite element mesh, which does not conform to the shape of the interface. Additionally, this approach can be utilised to model weak discontinuities, see Sect. 2.

Furthermore, weak discontinuities at interfaces, i.e. jumps in the strain field due to different material behaviour, can be fulfilled weakly in line with the ideas introduced in the mortar method. Firstly, Lagrange multipliers can be introduced to approximate the bond condition in a weak sense, see e.g. [8] for an application to fluid structure interaction. Secondly, Nitsche's method can be coupled with XFEM using a discontinuous enrichment functions as it is described in Sect. 3.

The different strategies are compared in Sect. 4. The optimisation approach including sensitivity analysis is briefly outlined in Sect. 5. Finally, Sect. 6 concludes this outline of an ongoing project on optimisation of multi-material structures with stress based constraints at the interface.

2 The Extended Finite Element Method

The displacement approximation in the extended finite element method (XFEM) consists of the standard continuous part and a discontinuous enhancement, which is based on special discontinuous enrichment functions. Two situations should be distinguished, i.e. strong discontinuities modelling cracks and weak discontinuities in case of interfaces. The later enrichment functions fulfil the continuity condition for displacements and model exactly the weak discontinuous physical behaviour in every point of the interface as outlined below. An alternative is described in Sect. 3.

In XFEM, the function spaces of the displacement approximations are enriched such that they either contain the solution or are closer to it. We discretise the body Ω into *nel* special finite elements. Let $\mathcal{N} = (n_1, n_2, .., n_m)$ be set of m nodes in the mesh. Depending on the nodal degrees of freedom (dofs) the set can be partitioned as $\mathcal{N} = \mathcal{N}^{\mathrm{std}} \cup \mathcal{N}^{\mathrm{enr}}$, where the set $\mathcal{N}^{\mathrm{enr}}$ consists of nodes corresponding to the cut elements. These so-called enriched nodes have additional degrees of freedom corresponding to the extra enrichment functions. The rest of nodes build the set

$\mathcal{N}^{\text{std}}$ with the standard set of unknowns. Similarly, the same split can be achieved for each element, i.e. $\mathcal{N}_e = \mathcal{N}_e^{\text{std}} \cup \mathcal{N}_e^{\text{enr}}$. The XFEM displacements $\mathbf{u}_h$ read

$$\mathbf{u}_h = \sum_{i \in \mathcal{N}_e} h^i \mathbf{u}_i + \sum_{j \in \mathcal{N}_e^{\text{enr}}} M^j \mathbf{a}_j, \tag{1}$$

where the shape functions $h^i := h^i(\mathbf{r})$ and the nodal displacements $\mathbf{u}_i$ contribute to the continuous part. The discontinuous part is given by the functions $M^j := M^j(\mathbf{r})$ and the extra degrees of freedom $\mathbf{a}_j$. Both functions depend on the coordinates $\mathbf{r}$ of the fixed parameter space. The strains ε_h and the stresses σ_h read

$$\varepsilon_h = \sum_{i \in \mathcal{N}_e} \mathbf{B}_{\text{std}}^i \mathbf{u}_i + \sum_{j \in \mathcal{N}_e^{\text{enr}}} \mathbf{B}_{\text{enr}}^j \mathbf{a}_j \quad \text{and} \quad \sigma_h = \mathbf{C}\varepsilon_h, \tag{2}$$

where $\mathbf{C}$ denotes the elasticity matrix. The index h is neglected from now on for notational simplicity. The B-Operators $\mathbf{B}_{\text{std}}^i$ and $\mathbf{B}_{\text{enr}}^j$ include the spatial derivatives of the standard h^i and special shape function M^j, respectively. For the sake of clarity we show the element matrices of unknowns, shape functions and B-operators

$$\mathbf{u}^e = [\mathbf{u}_i, \mathbf{a}_j], \quad \mathbf{H}^e = \left[\mathbf{H}_{\text{std}}^i, \mathbf{H}_{\text{enr}}^j\right] \quad \text{and} \quad \mathbf{B}^e = \left[\mathbf{B}_{\text{std}}^i, \mathbf{B}_{\text{enr}}^j\right]. \tag{3}$$

The special shape functions $M^j := M^j(\mathbf{r})$ can be defined by using the so-called *enrichment function* $\psi := \psi(\mathbf{r})$ and the standard shape function h^j in form of

$$M^j = h^j \psi. \tag{4}$$

We distinguish between conforming and non-conforming enrichment functions. In case of the conforming XFEM, the discontinuous physical behaviour is exactly represented by the enrichment function. Thus, ψ is discontinuous for cracks and ψ is continuous with discontinuous first derivative for interfaces. A so-called 'hat'-enrichment is recommended for interfaces, see [10], yielding

$$M^j = h^j \left(\sum_{i \in \mathcal{N}_e} h_i |\phi_i| - \left| \sum_{i \in \mathcal{N}_e} h_i \phi_j \right| \right). \tag{5}$$

Alternatively, the continuity condition is weakly fulfilled in all flavours of the non-conforming strategy. A XFEM-Lagrange approach is used e.g. in [8] in reference to the mortar-method using Lagrange-multipliers. Herein, the fluid-solid interaction problem is interpreted as an inf-sup problem for displacements and multiplier, i.e. the space of unknowns is extended even more. The accuracy of this approach has been verified, see [2] and Sect. 4.

3 Combination of Extended Finite Element and Nitsche Method

Inspired by Nitsche's method [11] we propose a novel non-conforming XFEM type. The XFEM-Nitsche approach is based on strong discontinuous enrichment functions, i.e. they are not conform to the physical discontinuous behaviour. The continuity of the solution across the interfaces is enforced weakly using Nitsche's method. The advantage of this method is, that both strong and weak discontinuities can be modelled using the same displacement approximation on the finite element level.

We consider a domain Ω with two non-overlapping subdomains Ω^- and Ω^+ and the common interface Γ_I of Ω^- and Ω^+ with normal vector $\mathbf{n} := \mathbf{n}_- = -\mathbf{n}_+$. The weak form of equilibrium for perfectly bonded subdomains reads $G = G_{\text{int}} - G_{\text{ext}} = 0$, where the external virtual work G_{ext} contains the prescribed traction forces $\bar{\mathbf{t}}$ on the Neumann boundary Γ_N of Ω and the body forces $\bar{\mathbf{b}}$ in the domain Ω, i.e.

$$G_{\text{int}} = \int_\Omega \delta\varepsilon : \sigma \, d\Omega \quad \text{and} \quad G_{\text{ext}} = \int_\Omega \delta\mathbf{u} \cdot \bar{\mathbf{b}} \, d\Omega + \int_{\Gamma_N} \delta\mathbf{u} \cdot \bar{\mathbf{t}} \, d\Gamma . \tag{6}$$

Furthermore, ε, σ and $\delta\mathbf{u}$ represent the strain and stress tensors and the test function.

The displacement jump $[\![\mathbf{u}]\!] = \mathbf{u}_+ - \mathbf{u}_-$ on the interface Γ_I must vanished for perfectly bonded interfaces. This constraint is enforced weakly using the Nitsche approach [11]. The steps to derive the corresponding weak form are outlined.

The standard weak form (6) is now formulated independently for each subdomain and both equations are added. Now, the interface is part of the each boundary, i.e. the traction forces $\sigma_- \cdot \mathbf{n}_-$ on Γ_I^- and $\sigma_+ \cdot \mathbf{n}_+$ on Γ_I^+ contribute to the weak form in form of a product with the virtual displacements $\delta\mathbf{u}_-$ and $\delta\mathbf{u}_+$, respectively. This contribution can be rewritten using the definition of the normal vector $\mathbf{n}$ yielding

$$G = G_{\text{int}} - G_{\text{ext}} + \int_{\Gamma_I} [\![\delta\mathbf{u} \cdot \sigma]\!] \cdot \mathbf{n} \, d\Gamma = 0. \tag{7}$$

The identity $[\![\delta\mathbf{u} \cdot \sigma]\!] \cdot \mathbf{n} = [\![\delta\mathbf{u}]\!] \cdot \{\sigma\} \cdot \mathbf{n} + \{\delta\mathbf{u}\} \cdot [\![\sigma]\!] \cdot \mathbf{n}$ and the strong continuity of the normal stress, i.e. $[\![\sigma]\!] \cdot \mathbf{n} = 0$ on Γ_I, yield the reformulation

$$G = G_{\text{int}} - G_{\text{ext}} + \int_{\Gamma_I} [\![\delta\mathbf{u}]\!] \cdot \{\sigma\} \cdot \mathbf{n} \, d\Gamma = 0 . \tag{8}$$

The mean stress is evaluated by using weighting factors dependent on geometry, i.e.

$$\{\sigma\} = \kappa^+ \sigma_+ + \kappa^- \sigma_- \quad \text{with} \quad \kappa^+ + \kappa^- = 1 . \tag{9}$$

The weak formulation (8) is neither symmetric nor stable. The first drawback can be repaired by adding $\int_{\Gamma_I} [\![\mathbf{u}]\!] \cdot \{\delta\sigma\} \cdot \mathbf{n} \, d\Gamma$ to enforce the symmetry. An additional stabilisation term is added using a fixed number $\theta > 0$ that must be chosen sufficiently large to ensure stability of the method. The value of θ depends on the material properties of the subdomain and on the size of the finite elements.

The weak form of Nitsche's method reads $G_{\text{Nitsche}} = G_{\text{int}} - G_{\text{ext}} + G_{\text{Interface}} = 0$, where $G_{\text{Interface}}$ represents the contribution of the interface, i.e.

$$G_{\text{Interface}} = \int_{\Gamma_I} [\![\delta\mathbf{u}]\!] \cdot \{\sigma\} \cdot \mathbf{n} \, d\Gamma + \int_{\Gamma_I} [\![\mathbf{u}]\!] \cdot \{\delta\sigma\} \cdot \mathbf{n} \, d\Gamma + \theta \int_{\Gamma_I} [\![\delta\mathbf{u}]\!] \cdot [\![\mathbf{u}]\!] \, d\Gamma .$$

This formulation represents the continuous bilinear form of the Nitsche method.

The domain is discretised independent of the material interfaces, which move over a background mesh. Thus, special attention must be devoted to the *ncut* elements, which are divided by the material interface. The discrete version of G_{Nitsche} reads

$$G_{\text{XFEM-Nitsche}} = \sum_{e=1}^{nel} G^e = \sum_{e=1}^{nel} G^e_{\text{int}} - \sum_{e=1}^{nel} G^e_{\text{ext}} + \sum_{e=1}^{ncut} G^e_{\text{dis}} = 0, \tag{10}$$

where G^e is the virtuell work on the element level. The approximation of the discrete weak formulation is based on the approximation of the quantities in each element, e.g. the displacements, the strains and the stresses as outlined in Sect. 2, i.e.

$$G^e_{\text{int}} = \delta\mathbf{u}^T \mathbf{k}^e \mathbf{u} \quad \text{with} \quad \mathbf{k}^e = \int_{\Omega^e} \mathbf{B}^T \mathbf{C} \mathbf{B}\, d\Omega \quad \text{and} \tag{11}$$

$$G^e_{\text{ext}} = \delta\mathbf{u}^T \mathbf{f}^e \quad \text{with} \quad \mathbf{f}^e = \int_{\Omega^e} \mathbf{H}^T \mathbf{b}\, d\Omega + \int_{\Gamma^e} \mathbf{H}^T \bar{\mathbf{t}}\, d\Omega. \tag{12}$$

The discontinuous XFEM displacement approximation in the cut elements is based on a *sign*-function, which is continuous inside the subdomains and enforces a jump in the displacements on the interface. The displacement approximations read

$$\mathbf{u}^+ = \sum_{i\in\mathcal{N}_e} h^i \mathbf{u}_i + \sum_{j\in\mathcal{N}_e^{\text{enr}}} M_+^j \mathbf{a}_j \quad \text{and} \quad \mathbf{u}^- = \sum_{i\in\mathcal{N}_e} h^i \mathbf{u}_i + \sum_{j\in\mathcal{N}_e^{\text{enr}}} M_-^j \mathbf{a}_j, \tag{13}$$

where M_+^j and M_-^j are given by shifted formulations in form of

$$M^j = \mathbf{h}_j \left(\psi(\phi) - \psi^j \right) \quad \text{with} \quad \psi(\phi) = \text{sign}(\phi) \quad \text{and} \quad \psi^j = \text{sign}(\phi_j), \tag{14}$$

which are evaluated in each subdomain Ω^- and Ω^+.

Thus, the displacement jump and the averaged stress within the cut elements read

$$[\![\mathbf{u}]\!] = \sum_{j\in\mathcal{N}_e^{\text{enr}}} \left(M_+^j - M_-^j \right) \mathbf{a}_j$$

$$\{\sigma\} = \mathbf{C} \left(\sum_{i\in\mathcal{N}_e} \mathbf{B}^i_{\text{std}} \mathbf{u}_i + \sum_{j\in\mathcal{N}_e^{\text{enr}}} \left(\kappa^+ \mathbf{B}^j_{\text{enr},+} + \kappa^- \mathbf{B}^j_{\text{enr},-} \right) \mathbf{a}_j \right).$$

For the sake of clarity we introduce the following matrices

$$[\![\mathbf{H}]\!] = \left[0, \left(M_+^j - M_-^j \right) \right] \quad \text{and} \quad \{\mathbf{H}\} = \left[h^i, \left(\kappa^+ M_+^j + \kappa^- M_-^j \right) \right]$$

$$[\![\mathbf{B}]\!] = \left[0, \left(\mathbf{B}^j_{\text{enr},+} - \mathbf{B}^j_{\text{enr},-} \right) \right] \quad \text{and} \quad \{\mathbf{B}\} = \left[\mathbf{B}^i, \left(\kappa^+ \mathbf{B}_+^j + \kappa^- \mathbf{B}_-^j \right) \right].$$

The element contribution G^e_{dis} of the interface part $G_{\text{Interface}}$ takes the form

$$G^e_{\text{dis}} = \delta\mathbf{u}^T \mathbf{k}^e_{\text{dis}} \mathbf{u}, \tag{15}$$

which is an additional conribution to the element stiffness matrix, i.e.

$$\mathbf{k}_{\text{dis}}^{e} = \int_{\Gamma_I^e} [\![\mathbf{H}]\!]^{T} \, \mathbf{C} \, \{\mathbf{B}\} \, \mathbf{n}^{T} \, d\Gamma + \int_{\Gamma_I^e} \{\mathbf{B}\}^{T} \, \mathbf{C} \, [\![\mathbf{H}]\!] \, \mathbf{n}^{T} \, d\Gamma + \int_{\Gamma_I^e} \theta^{e} \, [\![\mathbf{H}]\!]^{T} \, [\![\mathbf{H}]\!] \, d\Gamma \, .$$

This formulation for weak discontinuities can be directly transformed into a formulation for structures with strong discontinuities, e.g. cracks or delamination on the interface. In this case, the stiffness terms $G_{\text{Interface}}$ or G_{dis}^{e} or $\mathbf{k}_{\text{dis}}^{e}$ are neglected.

4   Comparison of Different XFEM Versions

The proposed Nitsche-type extended finite element method is compared with the non-conforming XFEM using Lagrange-multipliers. A quadrilateral domain Ω with circular inclusion and a sharp interface Γ_I is chosen, see Fig. 1. The shape of the inclusion is described using an implicit function. We assume different properties of the matrix matrial surrounding the inclusion and of the inclusion itself. Inside the inclusion we consider a soft material, defined by $E^{-} = 1$ and $v^{-} = 0.25$. Outside the inclusion a matrix material with $E^{+} = 10$ and $v^{+} = 0.3$ is assumed. The exact solution of this problem can be computed, see [13].

Three different version of the extended finite element method are considered, i.e. a conforming XFEM using an *abs-enrichment* and two non-conforming XFEM strategies, namely XFEM-Lagrange (XFEM-LMM) and XFEM-Nitsche. The convergence study is carried out using a set of different sizes of the element edges, i.e. $h = 0.1, 0.05, 0.025, 0.0125$. We only report the rate of convergence for the different computational techniques in terms of the relative L_2-norm of the error in the displacement field, see Fig. 2.

We obtain an optimal convergence rate by using the sign enrichment and enforcing the continuity condition either by Lagrange multiplier or by the Nitsche method.

Overall, the numerical results of these strategies differ hardly from the analytical solution. An ad-hoc application of any method, see XFEM-AbsEnr as an example, yields a less accurate solution. Other methods have also been compared, see [2] for first results of the accuracy of stresses.

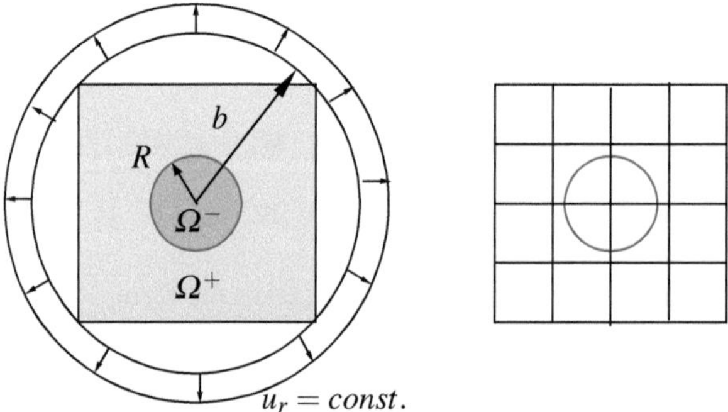

Fig. 1 Geometry and discretisation of plate with inclusion

Fig. 2 Convergence rate of
the applied techniques

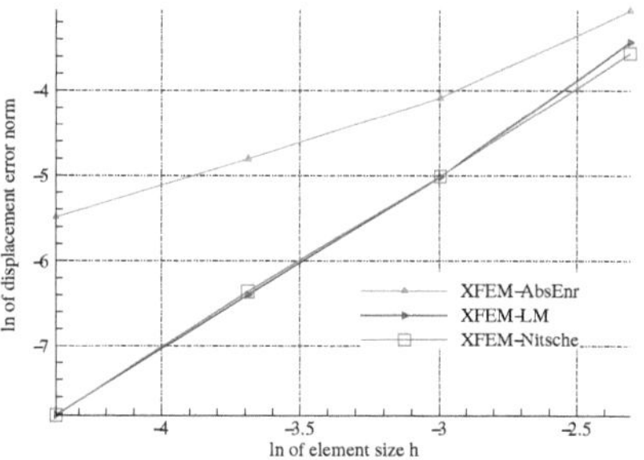

5 Structural Optimisation for Material Interfaces

The aim of structural optimisation is to find the optimal distribution (number and position) as well as the shape of inclusions in multi-material structures. The XFEM approach is based on a fixed mesh, where the edges of the elements are not conform to the material interface. To solve the optimisation problem we split the structural opimisation process into topology and shape optimisation parts.

Firstly, we use the topology optimisation to find the optimal distribution of the inclusions. Herein, we discretise the design space into finite elements and introduce control parameters for each element like in the SIMP method, see [5]. Alternatively, the *topological derivatives* can be applied to decrease the objective function, see [7] for the details concerning the minimisation of the overall compliance.

In topology optimisation, the material interfaces are explicitly represented by the edges of the finite elements. Thus, the goal of the second step is to find the optimal geometry of the inclusion. Here, a combination of XFEM as described above and a parametric representation of interfaces using implicit functions, e.g. superelliptic functions, is advocated. Thus, the updated design is described by updated parameters of the implicit function. Various researchers use the fixed grid strategy in structural optimisation. The implicit representation of holes with the level set method is a challenging task in topology and shape optimisation, see e.g. [1].

A sensitivity analysis of interfaces and of the mechanical properties on interfaces is a challenging task, which most often is solved only numerically. A discrete analytical sensitivity analysis of XFEM has been derived by the authors, see [12].

6 Conclusion

The extended finite element method has been reviewed briefly in order to motivate the formulation of a novel non-conforming version. The combination of extended finite elements using discontinuous enrichment functions and Nitsche's method to enforce weak continuity seems to be a promising approach. Especially the fact, that weak discontinuities can be easily transformed into strong discontinuities without altering the element formulation is an interesting observation.

Acknowledgements. The financial support of the Deutsche Forschungsgemeinschaft (DFG) under contract BA 1828/2-1 is greatfully acknowledged.

References

1. Allaire, G., Jouve, F., Toader, A.M.: Structural optimization using sensitivity analysis and a level set method. Journal of Computational Physics 194(1), 363–393 (2004)
2. Barthold, F.J., Rotthaus, M.: Remarks on interface models in structural analysis and sensitivity analysis. In: 7th World Congress on Structural and Multidisciplinary Optimization, Paper A0464, pp. 2459–2468 (2007)
3. Belgacem, F.: The mortar finite element method with lagrange multipliers. Numerische Mathematik 84, 173–197 (1999)
4. Belytschko, T., Black, T.: Elastic crack growth in finite elements with minimal remeshing. International Journal for Numerical Methods in Engineering 45(5), 601–620 (1999)
5. Bendsoe, M., Sigmund, O.: Topology Optimization: Theory, Methods and Applications. Springer, Heidelberg (2003)
6. Bernardi, C., Maday, Y., Patera, A.: A new nonconforming approach to domain decomposition: The mortar element method. In: Nonlinear partial differential equations and their applications, Longman, Paris. Pitman research Notes Mathematics, vol. 299, pp. 13–51 (1994)
7. Giusti, S., Novotny, A., Padra, C.: Topological sensitivity analysis of inclusion in two-dimensional linear elasticity. Engineering Analysis with Boundary Elements (2007) (accepted for publication)
8. Kölke, A.: Modellierung und Diskretisierung bewegter Diskontinuitäten in randgekoppelten Mehrfeldsystemen. Ph.D. thesis, TU Braunschweig (2005)
9. Korneev, V., Langer, U.: 22 Domain Decomposition Methods and Preconditioning. In: Encyclopedia of Computational Mechanics. John Wiley & Sons, Chichester (2004)
10. Moës, N., Cloirec, M., Cartraud, P., Remacle, J.: A computational approach to handle complex microstructure geometries. Comput. Methods Appl. Mech. Engrg. 192, 3163–3177 (2003)
11. Nitsche, J.: Über ein Variationsprinzip zur Lösung von Dirichlet-Problemen bei Verwendung von Teilräumen, die keinen Randbedingungen unterworfen sind. Abhandlungen aus dem mathematischen Seminar der Universität Hamburg 36, 9–15 (1970)
12. Rotthaus, M., Barthold, F.J.: Sensitivity analysis for interface problems using xfem and implicit shape representation. Int. J. Num. Meth. Eng. (2008) (Submitted for publication)
13. Sukumar, N., Chopp, D., Moës, N., Belytschko, T.: Modeling holes and inclusions by level sets in the extended finite-element method. Comput. Methods Appl. Mech. Engrg. 190, 6183–6200 (2001)
14. Wohlmuth, B.: A mortar finite element method using dual spaces for the lagrange multiplier. SIAM J. Numer. Anal. 38, 989–1012 (2000)
15. Wriggers, P.: Numerical Contact Mechanics. Springer, Heidelberg (2006)

One-Dimensional Microstructure Dynamics

Arkadi Berezovski, Jüri Engelbrecht, and Gérard A. Maugin

Abstract. Dispersive wave propagation in solids with microstructure is discussed in the small-strain approximation and in the one-dimensional setting. It is shown that the generalizations of wave equation based on continualizations of discrete systems as well as on homogenization methods can be recovered in the framework of the internal variable theory in the case of non-dissipative processes.

1 Introduction

It is well known that the propagation of linear elastic waves in a homogeneous medium is governed by the wave equation, which in the one-dimensional case reads

$$u_{tt} = c^2 u_{xx}, \tag{1}$$

where u is the displacement, c is the elastic wave speed.

If the medium is non-homogeneous, i.e. there is a certain microstructure, the wave propagation is accompanied by wave dispersion. Historically, the dispersion

Arkadi Berezovski

Centre for Nonlinear Studies, Institute of Cybernetics at Tallinn University of Technology, Akadeemia tee 21, 12618 Tallinn, Estonia

Tel.: +372-620-4264

Fax: +372-620-4151

e-mail: `Arkadi.Berezovski@cs.ioc.ee`

Jüri Engelbrecht

Centre for Nonlinear Studies, Institute of Cybernetics at Tallinn University of Technology, Akadeemia tee 21, 12618 Tallinn, Estonia

e-mail: `je@ioc.ee`

Gérard A. Maugin

Institut Jean Le Rond d'Alembert, Université Pierre et Marie Curie, UMR 7190, 4 Place Jussieu, 75252, Paris Cédex 05, France

e-mail: `gam@ccr.jussieu.fr`

J.-F. Ganghoffer, F. Pastrone (Eds.): Mech. of Microstru. Solids, LNACM 46, pp. 21–28.

effects were also investigated in the non-linear framework (cf. [3, 9], where many possible generalizations are outlined).

In the linear case, the simplest generalization of the wave equation was obtained by the homogenization of a periodically inhomogeneous medium [6, 17]

$$u_{tt} = c^2 u_{xx} + c^2 l^2 A_{22} u_{xxxx},\tag{2}$$

where l is an internal length parameter and A_{22} is a dimensionless coefficient.

Later on the same model was derived by a standard continualization procedure of the equations of motion for a system of discrete particles [16, 1].

Another modification of the wave equation was pointed out in [9] again on the basis of the continualization

$$u_{tt} = c^2 u_{xx} + l^2 A_{21} u_{xxtt},\tag{3}$$

and it was repeated later by means of homogenization methods [19, 5].

The combination of the two dispersion models gives (see also derivation in [4])

$$u_{tt} = c^2 u_{xx} + l^2 A_{21} u_{xxtt} + c^2 l^2 A_{22} u_{xxxx}.\tag{4}$$

This model was also derived from a discrete model by means a non-standard continualization procedure [13].

Recently, a "causal" model for dispersive wave propagation is proposed [14]

$$u_{tt} = c^2 u_{xx} + l^2 A_{21} u_{xxtt} + c^2 l^2 A_{22} u_{xxxx} - \frac{l^2}{c^2} A_{23} u_{tttt},\tag{5}$$

in order to avoid an infinite speed of propagation in the absence of higher-order time derivatives. The most general one-dimensional model based on the Mindlin theory of microstructure [15]

$$u_{tt} = \left(c^2 - c_A^2\right) u_{xx} - p^2 \left(u_{tt} - c^2 u_{xx}\right)_{tt} + p^2 c_1^2 \left(u_{tt} - c^2 u_{xx}\right)_{xx},\tag{6}$$

was discussed in [2]. Here c_A, c_1 and p are coefficients discussed in Section 4. It is clear that the last model includes all the previous ones.

Another approach to the description of microstructural effects is provided by the internal variable theory [12]. In this paper, we explain how the internal variable approach is generalized to the description of non-dissipative processes of linear dispersive wave propagation.

The paper is organized as follows. In the next Section we remind the reader of the canonical formulation of thermomechanics, which is applied even when no thermal effects are included. Then we present the internal variable theory specified to the non-dissipative processes. First we consider the case of one weakly non-local internal variable in the Section 3. In Section 4 we introduce an additional internal variable, which is necessary to obtain higher-order time derivatives.

2 Canonical Thermomechanics

The existence of the microstructure generally means that the medium is inhomogeneous. Therefore, we apply the canonical form of balance equations [8], where the inhomogeneities are treated in the most consistent way.

In the case of the thermoelastic conductors of heat, one-dimensional motion is governed by local balance laws for linear momentum and energy (no body forces)

$$\frac{\partial}{\partial t}(\rho_0 v) - \frac{\partial \sigma}{\partial x} = 0, \tag{7}$$

$$\frac{\partial}{\partial t}(\rho_0 v^2/2 + E) - \frac{\partial}{\partial x}(\sigma v - Q) = 0, \tag{8}$$

and by the second law of thermodynamics

$$\frac{\partial S}{\partial t} + \frac{\partial}{\partial x}(Q/\theta + K) \geq 0. \tag{9}$$

Here t is time, ρ_0 is the matter density, $v = u_t$ is the physical velocity, σ is the Cauchy stress, E is the internal energy per unit volume, S is the entropy per unit volume, θ is temperature, Q is the material heat flux, and the "extra entropy flux" K vanishes in most cases, but this is not a basic requirement.

Canonical form of the energy conservation. The canonical energy equation is obtained by introducing the free energy per unit volume $W := E - S\theta$ and taking into account the balance of linear momentum (7)

$$\frac{\partial(S\theta)}{\partial t} + \frac{\partial Q}{\partial x} = h^{int}, \quad h^{int} := \sigma\dot{\varepsilon} - \frac{\partial W}{\partial t}, \tag{10}$$

where the right-hand side of eqn. (10)$_1$ is formally an *internal heat source* [11].

In the case of non-zero extra entropy flux, the second law of thermodynamics gives

$$-\left(\frac{\partial W}{\partial t} + S\frac{\partial \theta}{\partial t}\right) + \sigma\dot{\varepsilon} + \frac{\partial}{\partial x}(\theta K) - (Q/\theta + K)\frac{\partial \theta}{\partial x} \geq 0, \tag{11}$$

where $\varepsilon = u_x$ is the one-dimensional strain measure. The dissipation inequality (11) can be also represented in the form

$$S\dot{\theta} + (Q/\theta + K)\frac{\partial \theta}{\partial x} \leq h^{int} + \frac{\partial}{\partial x}(\theta K). \tag{12}$$

Canonical (material) momentum conservation. Multiplying eqn. (7) by u_x we then check that eqn. (7) yields the following material balance of momentum (cf. [8])

$$\frac{\partial P}{\partial t} - \frac{\partial b}{\partial x} = f^{int} + f^{inh}, \tag{13}$$

where the *material momentum P*, the material *Eshelby stress b*, the material *inhomogeneity force* f^{inh}, and the material *internal force* f^{int} are defined by [8]

$$P := -\rho_0 u_t u_x, \quad b := -\left(\rho_0 v^2/2 - W + \sigma\varepsilon\right), \tag{14}$$

$$f^{inh} := \left(\frac{1}{2}v^2\right)\frac{\partial \rho_0}{\partial x} - \left.\frac{\partial W}{\partial x}\right|_{expl}, \quad f^{int} := \sigma u_{xx} - \left.\frac{\partial W}{\partial x}\right|_{impl}. \tag{15}$$

Here the subscript notations *expl* and *impl* mean, respectively, the derivative keeping the fields fixed (and thus extracting the explicit dependence on *x*), and taking the derivative only through the fields present in the function. The canonical equations for energy and momentum (10) and (13) are the most general expressions we can write down without a postulate of the full dependency of the free energy *W* [11].

3 Single Internal Variable

Up to now the microstructure was not specified. It can be prescribed by the specification of location, shape, and properties of inclusions, as, for example, in the case of periodic structures. If the microstructure is irregular, such a prescription is impossible. In the framework of the phenomenological continuum theory it is assumed that the influence of the microstructure on the overall macroscopic behavior can be taken into account by the introduction of an internal variable φ which we associate with the integral distributed effect of the microstructure. Then the free energy *W* is specified as the general sufficiently regular function of the strain, temperature, the internal variable, and its space gradient [11]

$$W = \overline{W}(u_x, \theta, \varphi, \varphi_x). \tag{16}$$

The *equations of state* (in a sense, mere definition of the partial derivatives of the free energy) are given by

$$\sigma = \frac{\partial \overline{W}}{\partial u_x}, \quad S = -\frac{\partial \overline{W}}{\partial \theta}, \quad \tau := -\frac{\partial \overline{W}}{\partial \varphi} \quad \eta := -\frac{\partial \overline{W}}{\partial \varphi_x}. \tag{17}$$

Following the scheme originally developed in [7] for materials with *diffusive* dissipative processes described by means of internal variables of state, we chose the non-zero extra entropy flux *K* in the form

$$K = -\theta^{-1}\eta\dot{\varphi}. \tag{18}$$

In this case, the "internal" material force and heat source each split in two terms according to

$$f^{int} = f^{th} + \widetilde{f}^{intr}, \quad h^{int} = h^{th} + \widetilde{h}^{intr}, \tag{19}$$

where the *thermal sources* and the "intrinsic" sources are given by [11]

$$f^{th} := S\frac{\partial}{\partial x}\theta, \quad h^{th} := S\dot{\theta}, \quad \widetilde{f}^{intr} := \widetilde{\tau}\frac{\partial \varphi}{\partial x}, \quad \widetilde{h}^{intr} := \widetilde{\tau}\dot{\varphi}, \tag{20}$$

so that we have the following consistent canonical equations of momentum and energy:

$$\frac{\partial P}{\partial t} - \frac{\partial \widetilde{b}}{\partial x} = f^{th} + \widetilde{f}^{intr}, \quad \frac{\partial (S\theta)}{\partial t} + \frac{\partial \widetilde{Q}}{\partial x} = h^{th} + \widetilde{h}^{intr}, \tag{21}$$

with dissipation

$$\Phi = \widetilde{h}^{intr} - \left(\frac{Q - \eta \dot{\varphi}}{\theta}\right) \frac{\partial \theta}{\partial x} \geq 0, \tag{22}$$

where we have introduced the new definitions [11]:

$$\widetilde{\tau} \equiv -\frac{\delta \overline{W}}{\delta \varphi} := -\left(\frac{\partial \overline{W}}{\partial \varphi} - \frac{\partial}{\partial x}\left(\frac{\partial \overline{W}}{\partial \varphi_x}\right)\right) = \tau - \eta_x,$$
$$\widetilde{b} = -\left(\rho_0 v^2/2 - W + \sigma u_x - \eta \varphi_x\right). \tag{23}$$

In this formulation the Eshelby stress $\widetilde{b}$ complies with its role of grasping all effects presenting gradients since the gradient of φ plays a role parallel to that of the deformation gradient u_x. The dissipation inequality (22) is automatically satisfied in the isothermal case if $\widetilde{\tau} = k\dot{\varphi}$ with $k \geq 0$ since

$$\Phi = k\dot{\varphi}^2 \geq 0. \tag{24}$$

The fully non-dissipative case corresponds to $k = 0$.

The simplest free energy dependence is a quadratic function [2]

$$\overline{W} = \frac{\rho_0 c^2}{2} u_x^2 + A\varphi u_x + \frac{1}{2}B\varphi^2 + \frac{1}{2}C\varphi_x^2. \tag{25}$$

Accordingly, the stress components $(17)_{3,4}$ are determined as follows:

$$\sigma = \frac{\partial \overline{W}}{\partial u_x} = \rho_0 c^2 u_x + A\varphi, \quad \eta = -\frac{\partial \overline{W}}{\partial \varphi_x} = -C\varphi_x, \tag{26}$$

and τ coincides with the interactive internal force

$$\tau = -\frac{\partial \overline{W}}{\partial \varphi} = -Au_x - B\varphi. \tag{27}$$

Consequently, the balance of linear momentum is rewritten as

$$u_{tt} = c^2 u_{xx} + \frac{A}{\rho_0}\varphi_x, \tag{28}$$

and the evolution equation for the internal variable in the fully non-dissipative case (with $k = 0$) reduces to

$$C\varphi_{xx} - Au_x - B\varphi = 0. \tag{29}$$

Evaluating the first space derivative of the internal variable from the last equation

$$\varphi_x = \frac{C}{B}\varphi_{xxx} - \frac{A}{B}u_{xx}, \tag{30}$$

and its third space derivative from eqn. (28)

26 A. Berezovski, J. Engelbrecht, and G.A. Maugin

$$\frac{A}{\rho_0}\varphi_{xxx} = \left(u_{tt} - c^2 u_{xx}\right)_{xx},\tag{31}$$

we will have, inserting the results into the balance of linear momentum

$$u_{tt} = c^2 u_{xx} + \frac{C}{B}\left(u_{tt} - c^2 u_{xx}\right)_{xx} - \frac{A^2}{\rho_0 B}u_{xx}.\tag{32}$$

It is clear that the obtained equation covers the first three models of the dispersive wave propagation mentioned in the Introduction. Equation (32) is the most general model for the dispersive wave motion provided by the standard internal variable theory. To go further, we need to introduce one more internal variable following [18].

4 Dual Internal Variables

Now we suppose that the free energy depends on the internal variables φ, ψ and their space derivatives $W = \overline{W}(u_x, \varphi, \varphi_x, \psi, \psi_x)$. Then the constitutive equations follow

$$\sigma := \frac{\partial \overline{W}}{\partial u_x}, \quad \tau := -\frac{\partial \overline{W}}{\partial \varphi}, \quad \eta := -\frac{\partial \overline{W}}{\partial \varphi_x}, \quad \xi := -\frac{\partial \overline{W}}{\partial \psi}, \quad \zeta := -\frac{\partial \overline{W}}{\partial \psi_x}.\tag{33}$$

We include into consideration the non-zero extra entropy flux similarly to the case of one internal variable

$$K = -\theta^{-1}\eta\dot{\varphi} - \theta^{-1}\zeta\dot{\xi}.\tag{34}$$

The generalization of the internal variable theory to the case of two internal variables is straightforward. The canonical equations of momentum and energy keep their form with appropriate modifications. It can be checked that in the considered case the intrinsic source terms are determined as follows

$$\widetilde{f}^{intr} := (\tau - \eta_x)\varphi_x + (\xi - \zeta_x)\psi_x, \quad \widetilde{h}^{intr} := (\tau - \eta_x)\dot{\varphi} + (\xi - \zeta_x)\dot{\psi}.\tag{35}$$

The latter means that the dissipation inequality in the isothermal case reduces to

$$\widetilde{h}^{intr} = (\tau - \eta_x)\dot{\varphi} + (\xi - \zeta_x)\dot{\psi} \geq 0.\tag{36}$$

It is easy to see that in the non-dissipative case ($\widetilde{h}^{intr} = 0$) the dissipation inequality (36) can by satisfied by the choice

$$\dot{\varphi} = L(\xi - \zeta_x), \qquad \dot{\psi} = -L(\tau - \eta_x),\tag{37}$$

where L is a coefficient. The latter two evolution equations express the duality between them: one internal variable is driven by another one and vice versa.

Keeping a quadratic function as the free energy dependence

$$\overline{W} = \frac{\rho_0 c^2}{2}u_x^2 + A\varphi u_x + \frac{1}{2}B\varphi^2 + \frac{1}{2}C\varphi_x^2 + \frac{1}{2}D\psi^2,\tag{38}$$

we include for simplicity only the contribution of the second internal variable itself. In this case, the stress components are the same as previously

$$\sigma = \frac{\partial \overline{W}}{\partial u_x} = \rho_0 c^2 u_x + A\varphi, \quad \eta = -\frac{\partial \overline{W}}{\partial \varphi_x} = -C\varphi_x, \quad \zeta = -\frac{\partial \overline{W}}{\partial \psi_x} = 0, \qquad (39)$$

as well as the interactive internal force τ

$$\tau = -\frac{\partial \overline{W}}{\partial \varphi} = -Au_x - B\varphi. \qquad (40)$$

The only new term is

$$\xi = -\frac{\partial \overline{W}}{\partial \psi} = -D\psi. \qquad (41)$$

It follows from eqns. (37), (39)$_3$, and (41) that

$$\dot{\varphi} = -LD\psi, \qquad (42)$$

i.e., the dual internal variable ψ is proportional to the time derivative of the primary internal variable $\dot{\varphi}$ in this particular case. It follows immediately from eqn. (42) that the evolution equation for the dual internal variable (37)$_2$ can be rewritten in terms of the primary one as the hyperbolic equation

$$\ddot{\varphi} = L^2 D(\tau - \eta_x). \qquad (43)$$

As a result, we can represent the equations of motion in the form, which includes only primary internal variable,

$$u_{tt} = c^2 u_{xx} + \frac{A}{\rho_0}\varphi_x, \quad I\varphi_{tt} = C\varphi_{xx} - Au_x - B\varphi, \qquad (44)$$

where $I = 1/(L^2 D)$ is an internal inertia measure.

In terms of stresses introduced by eqn. (33), the same system of equations is represented as

$$\rho_0 \frac{\partial^2 u}{\partial t^2} = \frac{\partial \sigma}{\partial x}, \quad I\frac{\partial^2 \varphi}{\partial t^2} = -\frac{\partial \eta}{\partial x} + \tau. \qquad (45)$$

It is worth to note that the same equations are derived in [4] based on different considerations.

Again, we can determine the first space derivative of the internal variable from eqn. (44)$_2$

$$\varphi_x = -\frac{I}{B}\varphi_{ttx} + \frac{C}{B}\varphi_{xxx} - \frac{A}{B}u_{xx}, \qquad (46)$$

and its third derivatives from eqn. (44)$_1$

$$\frac{A}{\rho_0}\varphi_{xxx} = \left(u_{tt} - c^2 u_{xx}\right)_{xx}, \quad \frac{A}{\rho_0}\varphi_{ttx} = \left(u_{tt} - c^2 u_{xx}\right)_{tt}. \qquad (47)$$

Inserting the results into the balance of linear momentum (44)$_1$, we obtain a more general equation

$$u_{tt} = c^2 u_{xx} + \frac{C}{B}\left(u_{tt} - c^2 u_{xx}\right)_{xx} - \frac{I}{B}\left(u_{tt} - c^2 u_{xx}\right)_{tt} - \frac{A^2}{\rho_0 B} u_{xx}. \tag{48}$$

It is easy to see, identifying $A^2 = c_A^2 B \rho_0, C = I c_1^2, B = I/p^2$, that the obtained equation is nothing else but the general model of the dispersive wave propagation (6).

Acknowledgements. Support of the Estonian Science Foundation (A.B. and J.E.) is gratefully acknowledged.

References

1. Askes, H., Suiker, A.S.J., Sluys, L.J.: A classification of higher-order strain-gradient models – linear analysis. Arch. Appl. Mech. 72, 171–188 (2002)
2. Engelbrecht, J., Berezovski, A., Pastrone, F., Braun, M.: Waves in microstructured materials and dispersion. Phil. Mag. 85, 4127–4141 (2005)
3. Engelbrecht, J., Braun, M.: Nonlinear waves in nonlocal media. Appl. Mech. Rev. 51, 475–488 (1998)
4. Engelbrecht, J., Cermelli, P., Pastrone, F.: Wave hierarchy in microstructured solids. In: Maugin, G.A. (ed.) Geometry, Continua and Microstructure, pp. 99–111. Hermann Publ., Paris (1999)
5. Fish, J., Chen, W., Nagai, G.: Non-local dispersive model for wave propagation in heterogeneous media: one-dimensional case. Int. J. Numer. Meth. Engng. 54, 331–346 (2002)
6. Kunin, I.A.: Theory of Microstructured Elastic Media, Nauka, Moscow (1975) (in Russian)
7. Maugin, G.A.: Internal variables and dissipative structures. J. Non-Equilib. Thermodyn. 15, 173–192 (1990)
8. Maugin, G.A.: Material Inhomogeneities in Elasticity. Chapman and Hall, London (1993)
9. Maugin, G.A.: Nonlinear Waves in Elastic Crystals. Oxford University Press, Oxford (1999)
10. Maugin, G.A.: Pseudo-plasticity and pseudo-inhomogeneity effects in materials mechanics. J. Elasticity 71, 81–103 (2003)
11. Maugin, G.A.: On the thermomechanics of continuous media with diffusion and/or weak nonlocality. Arch. Appl. Mech. 75, 723–738 (2006)
12. Maugin, G.A., Muschik, W.: Thermodynamics with internal variables. J. Non-Equilib. Thermodyn. 19, 217–249 (1994)
13. Metrikine, A.V., Askes, H.: One-dimensional dynamically consistent gradient elasticity models derived from a discrete microstructure–part 1: generic formulation. Eur. J. Mech. A/Solids 21, 555–572 (2002)
14. Metrikine, A.V.: On causality of the gradient elasticity models. J. Sound Vibr. 297, 727–742 (2006)
15. Mindlin, R.D.: Microstructure in linear elasticity. Arch. Rat. Mech. Anal. 16, 51–78 (1964)
16. Mühlhaus, H.B., Oka, F.: Dispersion and wave propagation in discrete and continuous models for granular materials. Int. J. Solids Struct. 33, 2841–2858 (1996)
17. Santosa, F., Symes, W.W.: A dispersive effective medium for wave propagation in periodic composites. SIAM J. Appl. Math. 51, 984–1005 (1991)
18. Ván, P., Berezovski, A., Engelbrecht, J.: Internal variables and dynamic degrees of freedom. J. Non-Equilib. Thermodyn. 33, 235–254 (2008)
19. Wang, Z.-P., Sun, C.T.: Modeling micro-inertia in heterogeneous materials under dynamic loading. Wave Motion 36, 473–485 (2002)

Strain Localization in Polyurethane Foams: Experiments and Theoretical Model

Giampiero Pampolini and Gianpietro Del Piero

Abstract. Strain localization has been observed in polyurethane foams subjected to confined compression up to 70% of deformation. This phenomenon is described by a one-dimensional model, in which the foam is represented as a chain of non-linear elastic springs with non-convex strain energy density, and localization is attributed to progressive phase transition.

Keywords: polyurethane foams, strain localization, non-convex strain energy, solid-solid phase transition.

1 Introduction

In the uniaxial compression of foam polymers, and of cellular materials in general, three regimes can be identified. The first regime is an almost linear response, the second is a stress plateau with large displacements occurring at almost constant load, and the last is a branch characterized by large stress increase under relatively moderate deformations. In the literature, two main approaches for modeling the mechanical response of foam polymers are used: a microstructural approach, reproducing directly the microstructure with a complex structure of beams, and a macroscopic approach, based on a full three-dimensional representation of the material

Giampiero Pampolini
Laboratoire de Méchanique et d'Acoustique, CNRS, 31 chemin Josep-Aiguier, 13402 Marseille, France
Università di Ferrara, Via Saragat 1, 44100 Ferrara, Italy
e-mail: `pampolini@lma.cnrs-mrs.fr;giampiero.pampolini@unife.it`

Gianpietro Del Piero
Università di Ferrara, Via Saragat 1, 44100 Ferrara, Italy
e-mail: `dlpgpt@unife.it`

J.-F. Ganghoffer, F. Pastrone (Eds.): Mech. of Microstru. Solids, LNACM 46, pp. 29–38.
springerlink.com

as a continuum. There has been a wide spread of models based on the first approach (e.g., [Gent and Thomas (1963)], [Gibson and Ashby (1997)], [Warren and Kraynik (1997)], [Gong et al. (2005)]). In particular, Gong et al. studied the response curves of open-cell foams, and [Gong and Kyriakides (2005)] proposed a model with a periodic structure given by a 14-sided polyhedron, called the Kelvin cell. The ligaments are modeled as shear-deformable extensional beams, and the basic material is assumed to be linearly elastic. In this approach, the numerical simulation is demanding from the computational point of view, due to the complexity of the assumed microstructure and to the difficulty of modeling the contact between beams in the post-buckling regime, see [Bardenhagen et al. (2005)]. In the continuum approach, the material is considered as homogeneous and hyperelastic, and subject to non-homogeneous deformations. Indeed, there is experimental evidence of strain localization, see for example [Lakes et al. (1993)], [Wang and Cuitinho (2002)]. In particular, strain localization along bands orthogonal to the loading direction was observed by Wang and Cuitio, and confirmed by experiments performed by the present authors on polyurethane foam specimens, subjected to axial compression up to 70%. The same three stages in the evolution of deformation have been observed by [Bastawros et al. (2000)] and [Bart-Smith et al. (1998)] in both open- and closed-cell aluminium alloy foams. In these cases, the deformation localizes in narrow bands, whose width is of the order of a cell diameter. These results support the interpretation, suggested in previous works [Wang and Cuitinho (2002)], [Gioia et al. (2001)], of the non-homogeneous deformation as a phase transition. Here we present a model in which the foam is represented as a chain of non-linear elastic springs with non-convex strain energy, and the localization is interpreted as a progressive phase change. The first part of the paper deals with the test procedure and the experimental results, and the second contains an outline of the theoretical model. A complete description of the model can be found in [Pampolini and Del Piero (2008)].

2 Experimental Tests

For the compression tests we used the load frame INSTRON 4467 located at the *Laboratorio di Materiali Polimerici* of the University of Ferrara, with a load cell of 500 N. The tests were made on commercial polyurethane foams in confined compression. For confinement we used a polystyrene box, clamped to the load frame. The polystyrene box is more rigid than the tested material and, being transparent, allowed us to control optically the evolution of the deformation. At the upper basis, the specimen was in contact with a steel plate fixed to the moving crosshead. Five specimens of dimensions $100 \times 100 \times 50$ mm were tested. The crosshead speed was chosen to be 1 mm/min, which we believe small enough to neglect all rate-dependent effects. The test

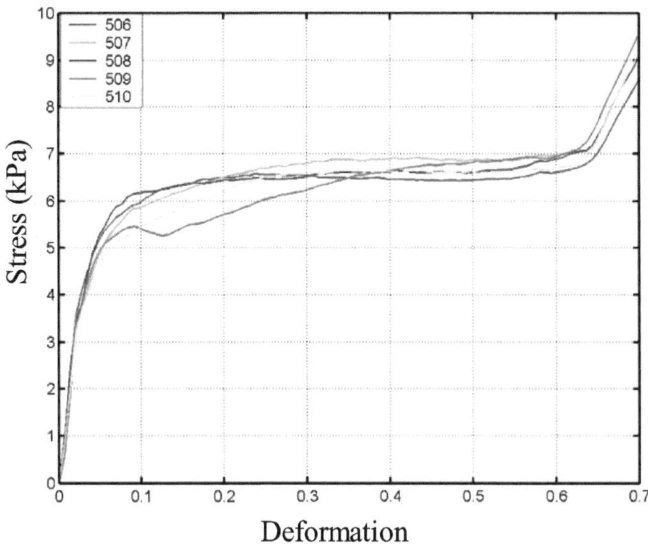

Fig. 1 Experimental curves of polyurethane foams in confined compression tests

was stopped at a relative elongation of 70%. As we will see later, this value is sufficiently large to include the significant portions of the response curves.

The stress-strain curves shown in Figure 1 reveal the characteristic behavior of polyurethane foams, and of cellular materials in general. From the

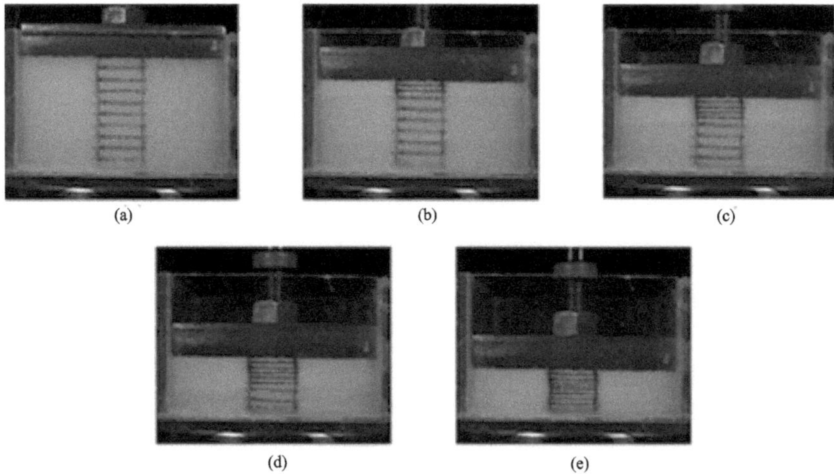

Fig. 2 Progressive deformation mechanism. Initial homogeneous deformation (a). Strain localization at the upper end of the specimen (b). Propagation to the underlying layers (c), (d). Back to homogeneous deformation (e)

figure, we notice that the plateau begins at a relative elongation of about 10% and ends at about 63%, and that the stress is close to 6,5 kPa. The slope of the second branch is approximatively 25 kPa. These values are close to those obtained by [Gioia et al. (2001)].

During the tests, the deformation was initially homogeneous (Figure 2a). Then a severe deformation occurred at the upper layer of the specimen (Figure 2b), and propagated to the underlying layers (Figures 2c, 2d). When all layers had reached the same deformation the specimen began again to deform homogeneously (Figure 2e). That the large deformation regime starts at one of the ends of the specimen can be attributed to a local boundary effect due to contact between steel plate and specimen. The same evolution mechanism was observed by [Wang and Cuitinho (2002)] in polyurethane low-density foams, and by [Bastawros et al. (2000)] and [Bart-Smith et al. (1998)] in aluminium alloy foams. In Wang and Cuitio's experiments, localization does not initiate at the upper part of the specimen as it does in our tests.

3 The Theoretical Model

3.1 Constitutive Assumptions

Let Ω_0 be the bounded open region of the three-dimensional space occupied by the body in the reference configuration, and let f be the deformation that maps the points X of Ω_0 into the points $x = f(X)$. Denote by Ω the region $f(\Omega_0)$ occupied by the deformed body, and by F the deformation gradient ∇f. Assume that the body is made of an isotropic hyperelastic material, with a strain energy density of the form

$$w(F) = \frac{1}{2}\alpha\, F \cdot F + \Gamma\,(\det F)\,. \tag{1}$$

where α is a positive material constant, and Γ is a function which takes the value $+\infty$ for both $\det F = 0$ and $\det F = +\infty$

$$\lim_{\det F \to 0} \Gamma\,(\det F) = \lim_{\det F \to \infty} \Gamma\,(\det F) = +\infty. \tag{2}$$

The gradient of w

$$S = w_F\,(F) = \alpha F + \det F\,\Gamma'\,(\det F)\,F^{-T}, \tag{3}$$

is the Piola-Kirchhoff stress tensor. If we impose that the reference configuration $F = I$ be stress-free, from the conditions $w(I) = w_F(I) = 0$ we get

$$\alpha = -\Gamma'(1), \quad \frac{3}{2}\alpha + \Gamma(1) = 0. \tag{4}$$

For the function Γ we assume the expression

$$\Gamma\left(\det F\right) = c\left(\det F\right)^n \left(\frac{1}{n+2}\left(\det F\right)^2 - \frac{1}{n}\right) - \mu \ln\left(\det F\right) + \frac{1}{2}\frac{\beta\sqrt{\pi}}{\sqrt{k}}\operatorname{erf}\left(\sqrt{k}\left(\det F - a\right)\right) + \gamma,$$

$$(5)$$

where $\alpha, c, n, \mu, \beta, k$ are positive constants, with $a \leq 1$, $\operatorname{erf}(\cdot)$ is the error function

$$\operatorname{erf}(x) = \int_0^x \frac{2}{\sqrt{\pi}} e^{-t^2}, \tag{6}$$

and γ is the constant determined by the condition $3\Gamma'(1) = 2\Gamma(1)$. The expression (5) of Γ consists of two parts. The first part is similar to that one proposed in [Ogden (1997)], and takes into account the long-range effects. The second part, the error function, provides a local effect at values of $\det F$ close to a. In the particular case of confined compression in the direction e, one has $F = I + (\lambda - 1)e \otimes e$. Then , for Γ as in (5), the energy density is

$$w(\lambda) = \frac{1}{2}\alpha\left(2 + \lambda^2\right) + \Gamma(\lambda), \tag{7}$$

and the normal component of S in the direction of loading is

$$\sigma = \left[\mu - \beta\exp\left(-k(1-a)^2\right)\right]\lambda + c\lambda^{n-1}\left(\lambda^2 - 1\right) - \mu\lambda^{-1} + \beta\exp\left(-k(\lambda - a)^2\right). \tag{8}$$

The energy has the non-convex form shown in Figure 3c, and the stress-strain curve has two ascending branches separated by a descending branch, as shown in Figure 3d.

3.2   *The Discrete Model*

Consider a chain of n springs connected in series as sketched in Figure 3a, each spring representing a layer of cells as shown in Figure 3b. We assume that all springs have the same strain energy w given by (7), and that, accordingly, the stress-strain curve is given by (8). The total energy of the system is the sum of the strain energies of the springs,

$$E\left(\varepsilon_1, \varepsilon_2, ..., \varepsilon_n\right) = \sum_{i=1}^{n} w(\varepsilon_i), \tag{9}$$

where ε_i is the elongation of the i-th spring. The bar is subjected to the *hard device* condition

$$\sum_{i=1}^{n} \varepsilon_i = n\varepsilon_0, \tag{10}$$

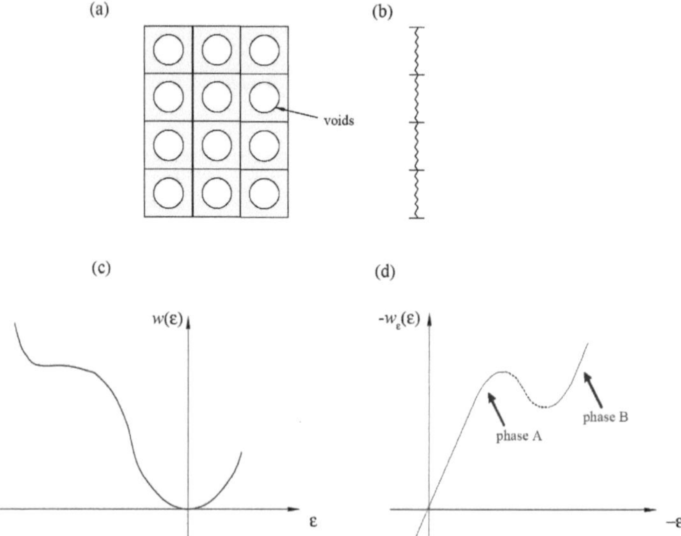

Fig. 3 Subdivision of the body into cell layers (a), representation of each layer as a non linear elastic spring (b) with non convex energy (c), and non-monotonic stress-strain curve (d)

where $n\varepsilon_0$ is the imposed displacement at the upper basis, the displacement at the lower basis being zero. By setting to zero the partial derivatives of the total energy, one gets the equilibrium condition

$$w'\left(\varepsilon_i\right) = w'\left(\varepsilon_n\right), \quad i = 1, 2, ..., n-1. \tag{11}$$

which implies that the force is the same in all springs. The common value of the force will be denoted by σ.

For sufficiently large n, a sufficient condition for stability is that all elongations ε_i lie on an ascending branch of the stress-strain curve, see [Del Piero and Truskinowsky (2008)] or [Puglisi and Truskinovsky (2000)]. Therefore, we confine our attention to the equilibrium configurations which satisfy this condition. For every such configuration, let m be the number of the springs whose elongations lie on the first ascending branch (phase A), so that $(n-m)$ is the number of the springs whose elongations lie on the second ascending branch (phase B). In particular, for $m = 0$ and for $m = n$ we have single-phase configurations, and for $0 < m < n$ we have two-phase configurations. In a single-phase configuration, all springs have the same elongation

$$\varepsilon_0 = \varepsilon_i, \quad i = 1, 2, ..., n, \tag{12}$$

and the global response curve (σ, ε_0) coincides with the response curve (σ, ε_i) of each spring. In a two-phase configuration, let $\bar{\sigma}$ be the force in the springs, and let $\bar{\varepsilon}_1$ and $\bar{\varepsilon}_2$ be the elongations corresponding to $\bar{\sigma}$ in the first and second ascending branch, respectively. Then the total elongation of the chain is

$$n\bar{\varepsilon}_0 = m\bar{\varepsilon}_1 + (n - m)\bar{\varepsilon}_2. \qquad (13)$$

By varying $\bar{\sigma}$, one can construct the equilibrium path $(\bar{\sigma}, \bar{\varepsilon}_0)$ corresponding to each given m. By dividing the interval $(\bar{\varepsilon}_1, \bar{\varepsilon}_2)$, into n equal parts, all equilibrium paths for different m are obtained. In Figure 4a, these paths are shown for a system of 4 springs.

If one assumes that the system evolves along equilibrium curves made of local energy minimizers, then the system, when loaded starting from the initial configuration, initially follows the first ascending branch. This branch ends when the stress reaches the critical value σ_{max} shown in Figure 4a. At this point, for further increasing deformation it is reasonable to assume that the system jumps to the closest stable branch, corresponding to the configuration with one spring in phase B and three springs in phase A. When this branch ends, the system jumps to the branch with two springs in phase B and so on, until all springs undergo the phase transition. At this point the system evolves following the second ascending branch, which corresponds to single-phase configurations with all springs in phase B. If we now increase the number of springs (Figure 4b), the number of the intermediate branches increases, and the amplitudes of the jumps at the end of the branches decrease. In this case when the stress reaches the critical value σ_{max}, the system follows a wavy, approximately horizontal line, successively assuming configurations with p springs in the highly deformed phase B and $n - p$ springs in the low-deformation phase A, with p gradually increasing from zero to n. At unloading the system exhibits a similar behavior, after reaching the critical

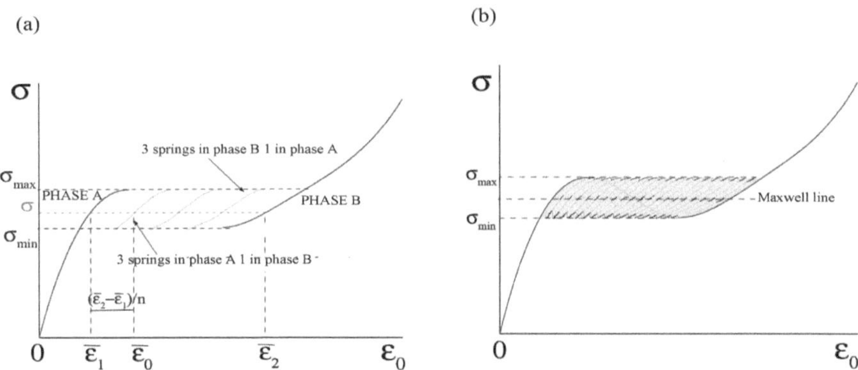

Fig. 4 Response curves for a system of 4 springs (a) and for a system of 20 springs (b)

stress σ_{min}. The model therefore predicts a hysteresis loop, which has indeed been observed experimentally [Gong and Kyriakides (2005)].

3.3 Comparison with the Experimental Data

In Figure 5 the model response is compared with the experimental results. The dotted line is the average loading curve, obtained as an average of the experimental curves shown in Figure 2. The solid line is the response curve of the model for a system of a sufficiently large number of springs with strain energy of the type (1), evaluated with the constitutive data reported in Table 1.

The model's response is close to the experimental curve in the first ascending branch and at large strains ($\varepsilon_0 > 0,3$). Instead of a

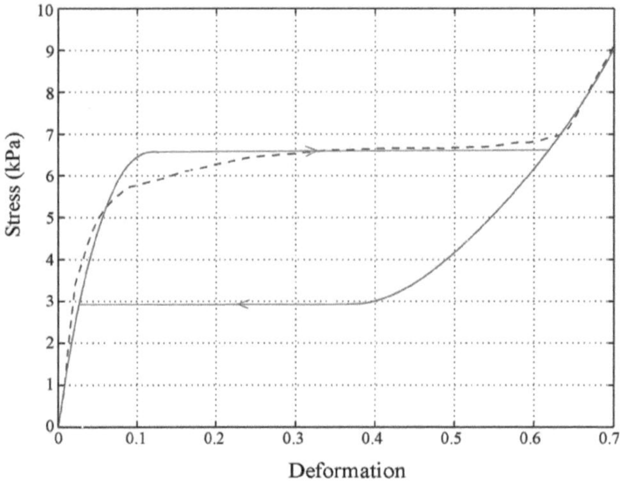

Fig. 5 Theoretical response curves and hysteresis loop at loading and unloading (solid lines), compared with experimental response at loading (dotted line)

Table 1 Values of the constitutive constants

		Material Constants
α		2.92 kPa
μ		2.92 kPa
c		69.9 kPa
m		6
β		3.5 kPa
k		18
a		0.72

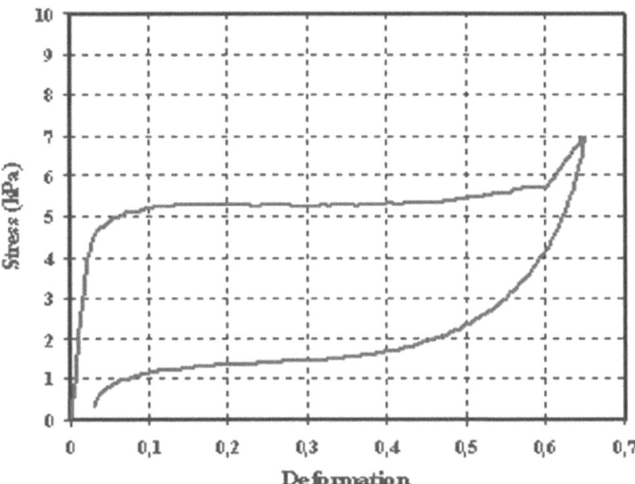

Fig. 6 Experimental response with hysteresis loop [Pampolini (2007)]

horizontal plateau, the experimental curve exhibits a branch with variable positive slope. This discrepancy has already been the object of investigation [Puglisi and Truskinovsky (2001)], [Marzano et al. (2003)]. In particular, Puglisi and Truskinovsky reproduce the positive slope by considering a chain of springs with the same ascending branches, but separated by different barriers. Finally, Figure 6 shows the experimental curve at unloading [Pampolini (2007)]. It confirms the presence of the hysteresis loop predicted by the proposed model.

Acknowledgement. This work was supported by the PRIN 2005 "Modelli Matematici per la Scienza dei Materiali" of the Italian Ministry for University and Research.

References

[Bardenhagen et al. (2005)] Bardenhagen, S.G., Brydon, A.D., Guilkey, J.E.: Insight into the physics of foam densification via numerical simulation. J. Mech. Phys. Solids 53, 597–617 (2005)

[Bart-Smith et al. (1998)] Bart-Smith, H., Bastawros, A.-F., Mumm, D.R., Evans, A.G., Sypeck, D.J., Wadley, H.N.G.: Compressive deformation and yielding mechanisms in cellular al alloys determined using x-ray tomography and surface strain mapping. Acta Materialia 46, 3583–3592 (1998)

[Bastawros et al. (2000)] Bastawros, A.-F., Bart-Smith, H., Evans, A.G.: Experimental analysis of deformation mechanism in a closed-cell aluminium alloy. J. Mech. Phys. Solids 48, 301–322 (2000)

[Del Piero and Truskinowsky (2008)] Del Piero, G., Truskinowsky, L.: Elastic bars with decohesions (2008) (in preparation)

[Gent and Thomas (1963)] Gent, A.N., Thomas, A.G.: Mechanics of foamed elastic materials. Rubber Chem. Technol. 36, 597–610 (1963)

[Gibson and Ashby (1997)] Gibson, L.J., Ashby, M.F.: Cellular Solids: Structure and Properties, 2nd edn. Cambridge University Press, Cambridge (1997)

[Gioia et al. (2001)] Gioia, G., Wang, Y., Cuitino, A.M.: The energetics of heterogeneous deformation in open-cell solid foams. Proc. R. Soc. London A 457, 1079–1096 (2001)

[Gong and Kyriakides (2005)] Gong, L., Kyriakides, S.: Compressive response of open-cell foams. part ii: Initiation and evolution of crushing. Int. J. Solids Struct. 42, 1381–1399 (2005)

[Gong et al. (2005)] Gong, L., Kyriakides, S., Jang, W.Y.: Compressive response of open-cell foams. part i: Morphology and elastic properties. Int. J. Solids Struct. 42, 1355–1379 (2005)

[Lakes et al. (1993)] Lakes, R., Rosakis, P., Ruina, A.: Microbuckling instability in elastomeric cellular solids. J. Mater. Sci. 28, 4667–4672 (1993)

[Marzano et al. (2003)] Marzano, S., Piccioni, M.D., Puglisi, G.: Un modello di isteresi per fili di leghe a memoria di forma. In: Proc. 16th AIMETA National Congress of Theoretical and Applied Mechanics (2003)

[Ogden (1997)] Ogden, R.W.: Non-Linear Elastic Deformations. Dover Publications, New York (1997)

[Pampolini (2007)] Pampolini, G.: Experimental tests and theoretical model for strain localization and cyclic damage of polyurethane foam cylinders in uniaxial compression. In: Communication at the VI Meeting Unilateral Problems in Structural Analysis, Siracusa (2007)

[Pampolini and Del Piero (2008)] Pampolini, G., Del Piero, G.: Strain localization in polyurethane foams. Experiments and theoretical model (2008) (in preparation)

[Puglisi and Truskinovsky (2000)] Puglisi, G., Truskinovsky, L.: Mechanics of a discrete chain with bi-stable elements. J. Mech. Phys. Solids 48, 1–27 (2000)

[Puglisi and Truskinovsky (2001)] Puglisi, G., Truskinovsky, L.: Hardening and hysteresis in transformational plasticity. In: Proc. 15th AIMETA National Congress of Theoretical and Applied Mechanics (2001)

[Wang and Cuitinho (2002)] Wang, Y., Cuitinho, Y.: Full-field measurements of heterogeneous deformation patterns on polymeric foams using digital image correlation. Int. J. Solids Struct. 39, 3777–3796 (2002)

[Warren and Kraynik (1997)] Warren, W.E., Kraynik, A.M.: Linear elastic behavior of a low-density kelvin foam with open cells. ASME J. Appl. Mech. 64, 787–793 (1997)

A Finite Element Approach of the Behaviour of Woven Materials at Microscopic Scale

Damien Durville

Abstract. A finite element simulation of the mechanical behaviour of woven textile materials at the scale of individual fibers is proposed in this paper. The aim of the simulation is to understand and identify phenomena involved at different scales in such materials. The approach considers small patches of woven textile materials as collections of fibers. Fibers are modelled by 3D beam elements, and contact-friction interactions are considered between them. An original method for the detection of contacts, and the use of efficient algorithms to solve the nonlinearities of the problem, allow to handle patches made of few hundreds of fibers. The computation of the unknown initial configuration of the woven structure is carried out by simulating the weaving process. Various loading cases can then be applied to the studied patches to identify their mechanical characteristics. To approach the mesoscopic behaviour of yarns, 3D strains are calculated at the scale of yarns, as post-processing. These strains display strong inhomogeinities, which raises the question of using continuous models at the scale of yarns.

1   Introduction

The mechanics of woven materials can be adressed at three different scales : the macroscopic scale relevant to pieces of fabric, the mesoscopic scale related to yarns, and the microscopic scale concerning fibers inside yarns. Different approaches are available to handle the problem of the identification and characterization of the complex mechanical behaviour of these materials. Some developments are specially dedicated to the characterization of the geometry of yarns in the woven structure ([6, 5]). Geometries of yarns obtained by these methods can then be meshed and classical finite element codes can be employed to compute the mechanical response of textile

Damien Durville
LMSSMat - Ecole Centrale Paris , Grande Voie des Vignes,
92290 Chatenay-Malabry, France
e-mail: `damien.durville@ecp.fr`

J.-F. Ganghoffer, F. Pastrone (Eds.): Mech. of Microstru. Solids, LNACM 46, pp. 39–46.
springerlink.com

or textile composite structures. Other approaches concentrate on the finite element simulation at the scale of yarns, representing yarns by means of 3D elements with appropriate constitutive laws [1]. Going down to the scale of fibers requires a description of the geometry of fibers in the initial configuration. Such a geometry is a priori unknown, and therefore needs to be computed, for example by simulating the weaving process. Such an approach has been proposed using a finite element simulation code based on an explicit solver [4]. The method we present here is based on a simulation code specially developped to handle the mechanical behaviour of entagled media [2], and makes use of an implicit solver. This simulation code is able to handle small patches of fabric made of few hundreds of fibers in order to identify their mechanical behaviour [3]. After recalling some basic features of the method, the way the weaving process is simulated to compute the initial configuration of the woven structure is presented, and results for two different weave patterns are given. Classical loading tests are then applied to these patches. The last part is dedicated to the calculation of 3D strains at the scale of yarns by means of meshes generated as post-processing. The results show strong inhomegeneities of strains at the scale of yarns, induced by localized displacements taking place between fibers.

2   Purposes of the Simulation at Microcopic Scale

The needs related to the modelling and simulation of textile materials are diversified. Formability studies require an identification of the global macroscopic behaviour of fabrics, whereas the focus will be put on the local behaviour of individual fibers if damage and fiber breaking phenomena are to be investigated.

The coupling between phenomena at different scales makes the study of such materials complex. The two main mechanisms ruling the behaviour of textile materials, namely the interweaving of yarns constituting the fabric, and the entanglement of fibers within the yarns, occur at different scales but must be considered simultaneaously. One possible strategy to solve the problem at the macroscopic scale is to formulate a mesoscopic model for yarns accounting for the behaviour of the bundles of fibers making up the yarns. However, at the scale of yarns, one important issue is to determine whether phenomena between discrete fibers within yarns may be approached by continuous models.

To explore these questions, we propose to adress the global problem at the microscopic scale of individual fibers, by considering small patches of fabrics as assemblies of individual fibers, arranged in bundles, and developping between them contact-friction interactions. By this way, models at intermediate scales are no longer required, and the only mechanical characteristics to be known are those of individual fibers. As a downside, the initial geometry of fibers in the woven structure can not be predicted a priori. It is therefore necessary to compute this initial geometry by simulating the weaving process.

3 Main Features of the Simulation Code

3.1 A Kinematically Enriched Beam Model for Fibers

The beam elements used to take into account fibers in the simulation are based on an enriched kinematical model that describes kinematics of each cross-section by the means of three vector fields : one for the translation of the centroid, and two to represent planar and linear deformations of the cross-section. This model, characterized by nine degrees of freedom, allows to calculate full 3D strain tensors, accounting for deformations of cross-sections.

3.2 Modelling of Contact between Fibers

An original method [2, 3] has been developped to detect quicky and accurately the numerous contacts taking place within a collection of fibers submitted to large deformations. This method is based on the determination of contact elements made of pairs of material particles that are predicted to enter into contact. The determination of these elements relies on the construction of intermediate geometries in any region where two parts of beam are sufficiently close to each other. These geometries are defined as the average of the two close parts of line. Normal directions to these geometries are used as contact search directions to define material particles of contact elements. By this way, the process of determination of contact elements is symmetrical with respect to the two considered beams –which is not the case when the normal direction to only one structure is chosen as contact search direction.

3.3 Rigid Bodies for the Driving of Boundary Conditions

To simulate the various conditions corresponding to the initial weaving process or the different loading tests, a versatile driving of boundary considitions is required. For this purpose, rigid bodies are attached to each end of yarn, and to each side of the patch. These rigid bodies can be driven either by displacements or by forces, applied to their degrees of freedom in both translation and rotation.

4 Computation of the Initial Configuration: Simulation of the Weaving Process

4.1 Modelling of the Weaving Process

The simulation of the weaving process is necessary to determine the unknown initial configuration of the woven structure. The weave pattern specifies which yarn must

above or below the other at each crossing. The basic idea to simulate the weaving
process is to make these conditions progressively fulfiled.

To do so, we start from a flat configuration where all yarns are straight and pen-
etrate each other at crossings. Then, for few steps, for any penetration detected between two fibers belonging to crossing yarns, we take as normal direction of contact
a vertical direction oriented according to the local crossing order prescribed by the
weave pattern. This process is applied until yarns are completely separated at crossings. Then, in a second stage, classical contact directions, depending only on the
local geometry of fibers, are considered, while forces and displacements applied
on the sides of the patch are progressively relaxed. At the end of this process, an
equilibrium configuration of the woven structure after weaving is obtained.

4.2 *Application to the Studied Patches*

Two patches of fabric, made with the same yarns, but according to two different
weave patterns –a plain weave and a twill weave–, have been considered for the

Table 1 Characteristics of the studied patches

Nb. of fibers weft yarns	44
Nb. of fibers warp yarns	24
Nb. of weft yarns	6
Nb. of warp yarns	6
Total nb. of fibers	408
Nb. of nodes	35.000
Nb. of dofs	300.000
Nb. of contact elements	≈ 80.000

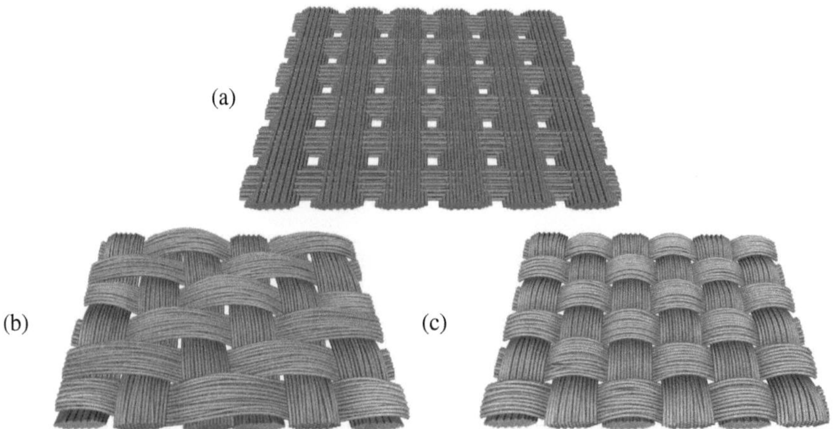

Fig. 1 Simulation of the weaving process : configuration before weaving (a), computed configuration for plain weave (b) and twill weave (c)

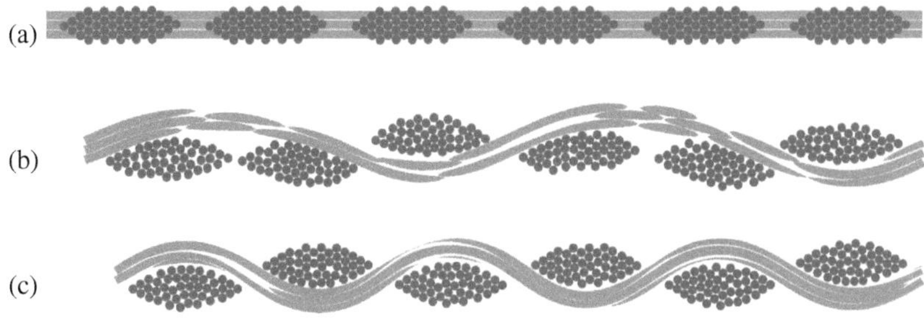

Fig. 2 Cuts of configuration before weaving (a), and of computed configuration for a plain weave (b) and a twill weave (c) patches

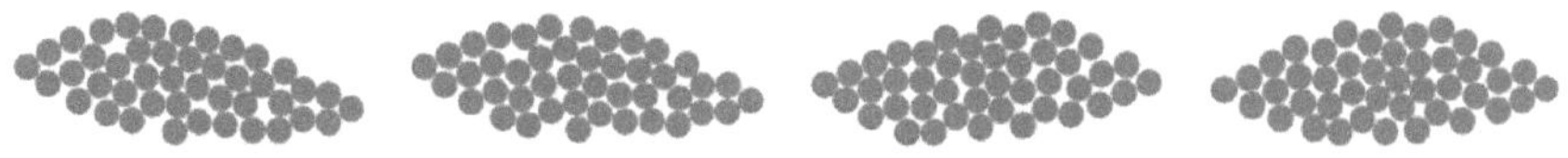

Fig. 3 Variation of cross-section shapes along a warp yarn for twill weave

results presented here. The main features characterizing these patches are given in Table 1. Each patch is made of 408 fibers, and about 80.000 contact elements are considered in the simulation.

Computed configurations for the two weavings are shown on Fig. 1. Cuts of the two computed initial configurations and details of these cuts (Fig. 2) show the re-arrangement of fibers and typical shapes of cross-sections obtained by the simulation. Whereas for plain weaves shapes of yarn cross-sections are predominantly lenticular, for twill weave cross-sections are more complex and vary along the yarn (Fig. 3).

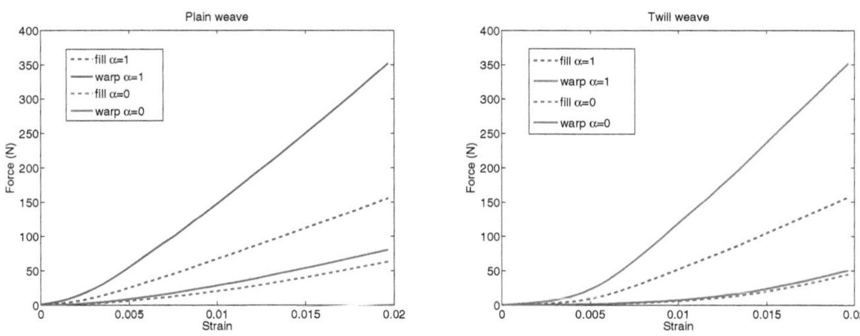

Fig. 4 Force/strain curves for biaxial tension tests for plain weave and twill weave

4.3 Application of Loading Tests

Once the initial configuration has been computed, various loading cases can be simulated by applying appropriate loadings (displacements or forces) to rigid bodies attached to each side of the patches. Biaxial tension loadings are applied up to a 2% extension in the warp direction, with a ratio α, taken successively equal to 0 and 1, for the extension in the weft direction. Typical J-shape force/displacement curves are obtained (Fig. 4). Nonlinear effects at the start of the curves are related to the compaction of cross-sections as the tensile force increases.

5   Strain Measures at the Scale of Yarns

The formulation of mesoscopic models to represent the behaviour of yarns requires to define measures for both strains and stresses at the scale of yarn. The definition of a stress measure is difficult because stresses within yarns are of two different kinds: continuous stresses inside yarns, and discrete contact-friction interactions between fibers.

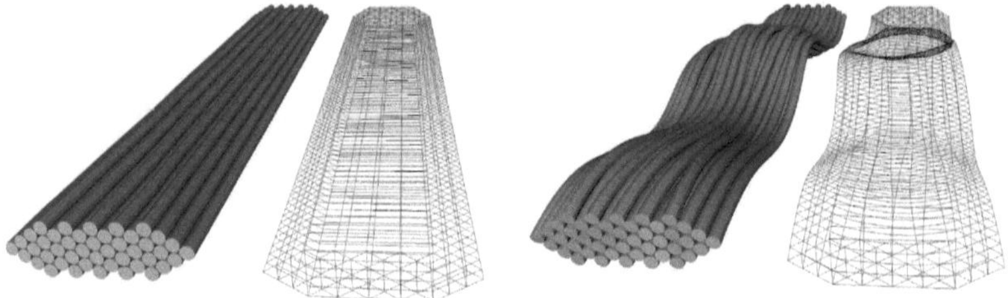

Fig. 5 Fibers of a yarn, and corresponding 3D meshes in two different configurations

As a first step, the definition of a strain measure at the scale of yarn is easier. To approximate such a quantity, we propose to consider each yarn as a continuum, and to compute, as post-processing, Green-Lagrange strain tensors using a 3D finite element mesh based on the nodes defined on fibers (Fig.5). Once the mesh connectivity has been defined for the initial configuration, nodal displacements and strain tensors can be easily derived. It is also possible to consider strains generated by displacements only between two given loading increments.

The plotting of horizontal strains due to the initial forming (Fig. 6) shows strong inhomogeneities. Zones looking like diagonal shear bands can be observed on the cuts of strains in both horizontal and vertical directions (Fig. 7 and 8).

Similar inhomogeneities can be observed for strains induced by the equibiaxial tension (Fig. 9). In this case, the axial strain is twice higher on the sides of the yarn than in the center. The appearance of shear bands shows a rearrangement of fibers in the center of the yarn. This rearrangement reduces the heigth of the cross-section, allowing fibers to undergo lower extensions.

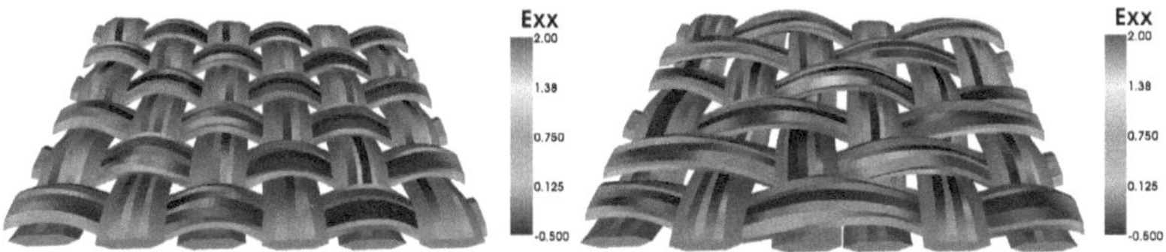

Fig. 6 Horizontal strains in yarns generated by the simulation of the weaving process

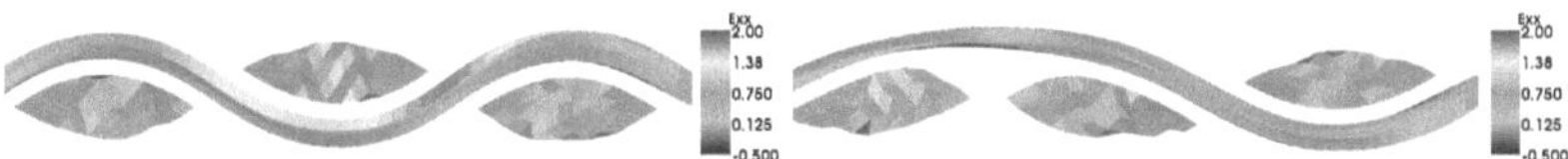

Fig. 7 Cuts of horizontal strains in yarns generated by the simulation of the weaving process

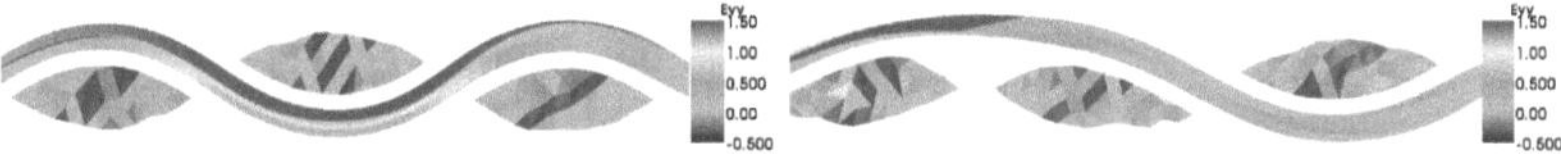

Fig. 8 Cuts of vertical strains in yarns generated by the simulation of the weaving process

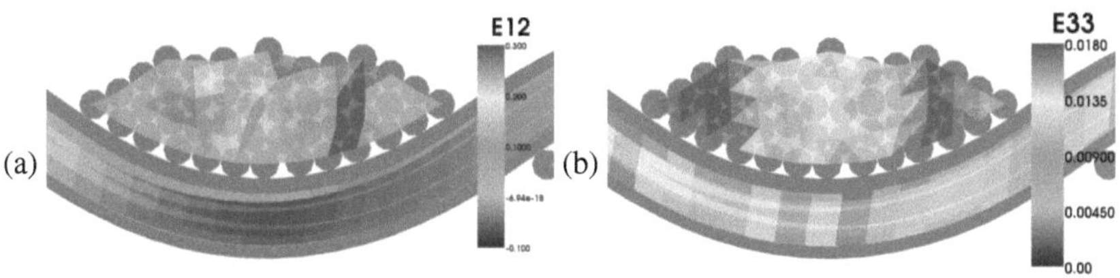

Fig. 9 Shear strains (a) and axial strains (b) in yarns generated by equibiaxial tension

Inhomogeneities revealed by this post-processing are of first importance since they raise the question of using continuous models to describe the behaviour of yarns.

6 Conclusion

Models and algorithms developped for the simulation of the mechanical behaviour of entangled media can be applied for the modelling of woven structures. Patches of fabric made of few hundreds of fibers can be considered by the model. The simulation of the weaving process provide accurate geometrical descriptions of both yarns trajectories and varying shapes of yarns cross-sections. The mechanical response of the fabric can then be characterized by simulating typical loading tests. The computation of 3D strains at the scale of yarns reveals strong inhomogeneities and raises the question of the validity of considering yarns as continuums.

References

1. Boisse, P., Zouari, B., Gasser, A.: A mesoscopic approach for the simulation of woven fibre composite forming. Composites Science and Technology 65(3-4), 429–436 (2005)
2. Durville, D.: Numerical simulation of entangled materials mechanical properties. Journal of Materials Science 40(22), 5941–5948 (2005)
3. Durville, D.: Finite element simulation of the mechanical behaviour of textile composites at the mesoscopic scale of individual fibers. In: Recent Advances in Textile Membranes and Inflatable Structures, pp. 15–34. Springer, Heidelberg (2008)
4. Finckh, H.: Numerical simulation of mechanical properties of fabrics - weaving / numerische simulation der mechanischen eigenschaften textiler flächengebilde - gewebeherstellung. In: Proceedings of the German 3rd LS-DYNA Forum 2004, Bamberg, Germany (2004)
5. Lomov, S., Ivanov, D., Verpoest, I., Zako, M., Kurashiki, T., Nakai, H., Hirosawa, S.: Meso-fe modelling of textile composites: Road-map, data flow and algorithms. Composites Science and Technology 67, 1870–1891 (2007)
6. Verpoest, I., Lomov, S.: Virtual textile composites software wisetex: Integration with micro-mechanical, permeability and structural analysis. Composites Science and Technology 65, 2563–2574 (2005)

Nonlinear Problems of Fibre Reinforced Soft Shells

Polina Dyatlova

Abstract. Interaction of soft shells with solids covered by them is investigated by finite elements method. The shells supposed to be made of thin membranes which are much more compliant in respect of stretching then the fibre nets by which these membranes are reinforced. Problems of numerical procedure's convergence are discussed.

1 Introduction

The development of methods for macroscopic analysis of stress-strained state of soft shells, i.e. the shells that don't resist bending forces, is necessary for improving many processes and products of various branches of industry. In this work the variational method for investigation of such shells reinforced by elastic filaments is being developed. Realization of this method is based on some given in [1] formulas for calculation of density of a strain's potential energy of the shell's material. By these formulas the density is determined as function of the shell's strain measures. The relative elongations of lines of material coordinates and changes of the angles between these lines are chosen as these measures. The shells are supposed to be infinitely thin in a macroscopic sense.

The general macroscopic theory of such shells has been developed [2], [3]. The calculations of the density of the strain's potential energy carried out with taking into account the properties of textile microstructures of the shells. As a rule such calculations represent significant difficulties. It is because general methods for explication of interconnections between strains of microstructures of the shells and the shell's macroscopic strains aren't developed.

The potential energy of the whole shell is calculated as integral of the density of strains' energy. To investigate by variational methods the interaction of the

Polina Dyatlova
dyatlovaspb@hotbox.ru

J.-F. Ganghoffer, F. Pastrone (Eds.): Mech. of Microstru. Solids, LNACM 46, pp. 47–56.
springerlink.com   © Springer-Verlag Berlin Heidelberg 2009

shell with some solid enveloped by it this integral must be considered for an arbitrary position of the shell on the solid's surface. The actual position of the shell is the one, for which this integral, i.e. the potential energy of the whole shell, has a minimal value.

2 Reduction of the Problem to Minimization of Potential Strain Energy of the Shell

Suppose that at some non-stressed state the form of the shell (see Fig. 1) in cylindrical coordinates R, φ, ζ is defined by equation

$$R = R_0(\varphi, \zeta). \tag{1}$$

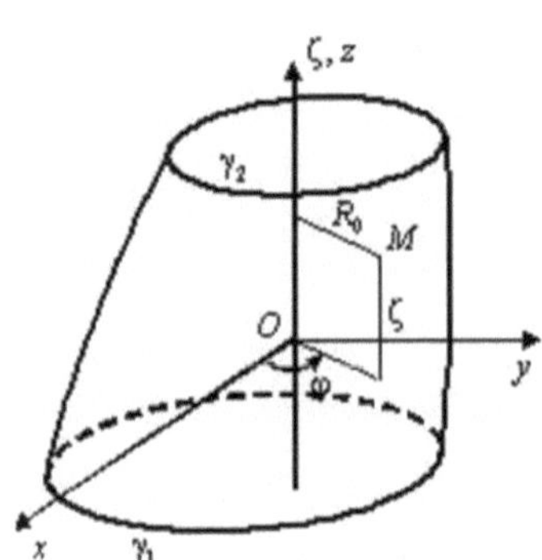

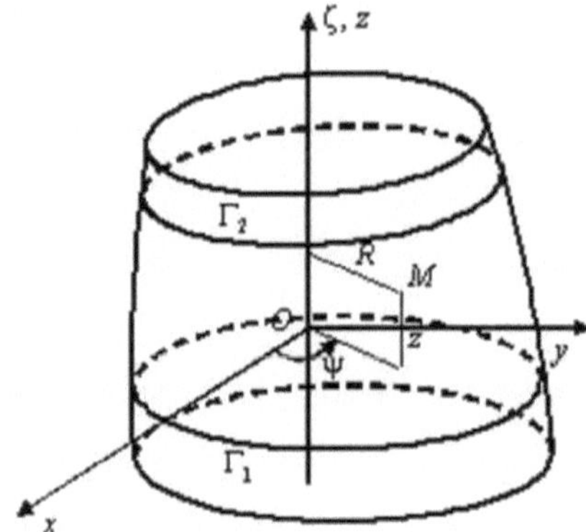

Fig. 1 The shell in unstressed state　　　**Fig. 2** The shell on the solid's surface

Further on variables φ and ζ are considered as Lagrange coordinates on the shell. The surface of some solid on which the shell will be pulled on is defined in cylindrical coordinates R, ψ, z by equation

$$R = R(\mu, \psi, z), \tag{2}$$

where axis z coincide with axis ζ (see Fig. 2), μ is some artificially introduced parameter variation of which leads to changing of solids form.

The radius-vector of any particle M of the shell when it is pulled on the solid's surface can be presented as follows

$$\mathbf{r} = \mathbf{i}(R_0(\varphi, \zeta) + \rho(\mu, \varphi, \zeta)) \cos(\varphi + \theta(\mu, \varphi, \zeta)) + \tag{3}$$
$$+ \mathbf{j}(R_0(\varphi, \zeta) + \rho(\mu, \varphi, \zeta)) \sin(\varphi + \theta(\mu, \varphi, \zeta)) + \mathbf{k}(\zeta + w(\mu, \varphi, \zeta)),$$

where $\theta(\mu, \varphi, \zeta)$ is increment of angular coordinate of the particle M, $\rho(\mu, \varphi, \zeta)$ and $w(\mu, \varphi, \zeta)$ are its radial and vertical displacements respectively; $\mathbf{i}$, $\mathbf{j}$, $\mathbf{k}$ – orts of axes x, y and z respectively.

It is obvious that following equalities are true

$$R(\mu, \varphi + \theta(\mu, \varphi, \zeta), \zeta + w(\mu, \varphi, \zeta)) = R_0(\varphi, \zeta) + \rho(\mu, \varphi, \zeta), \qquad (4)$$
$$\psi = \varphi + \theta(\mu, \varphi, \zeta),$$
$$z = \zeta + w(\mu, \varphi, \zeta),$$

To describe the deformations of the shell arising because it is put on the surface of the solid, we shall write down expressions for its metric coefficients corresponding, firstly, to its initial non-stressed state, and, secondly, to its final, deformed state. Accordingly to $(1) - (4)$, these coefficients are given by equalities

$$g_{11}^0 = \left(\frac{\partial \mathbf{r}}{\partial \varphi}\right)^2 = \left(\frac{\partial R_0}{\partial \varphi}\right)^2 + R_0^2, \qquad (5)$$

$$g_{22}^0 = \left(\frac{\partial \mathbf{r}}{\partial \zeta}\right)^2 = \left(\frac{\partial R_0}{\partial \zeta}\right)^2 + 1,$$

$$g_{12}^0 = g_{21}^0 = \frac{\partial \mathbf{r}}{\partial \varphi}\frac{\partial \mathbf{r}}{\partial \zeta} = \frac{\partial R_0}{\partial \varphi}\frac{\partial R_0}{\partial \zeta}$$

and

$$g_{11} = \left(\frac{\partial R_0}{\partial \varphi} + \frac{\partial \rho}{\partial \varphi}\right)^2 + (R_0 + \rho)^2\left(1 + \frac{\partial \theta}{\partial \varphi}\right)^2 + \left(\frac{\partial w}{\partial \varphi}\right)^2, \qquad (6)$$

$$g_{22} = \left(\frac{\partial R_0}{\partial \zeta} + \frac{\partial \rho}{\partial \zeta}\right)^2 + (R_0 + \rho)^2\left(\frac{\partial \theta}{\partial \zeta}\right)^2 + \left(1 + \frac{\partial w}{\partial \zeta}\right)^2,$$

$$g_{12} = g_{21} = \left(\frac{\partial R_0}{\partial \varphi} + \frac{\partial \rho}{\partial \varphi}\right)\left(\frac{\partial R_0}{\partial \zeta} + \frac{\partial \rho}{\partial \zeta}\right) +$$
$$+ (R_0 + \rho)^2\left(1 + \frac{\partial \theta}{\partial \varphi}\right)\left(\frac{\partial \theta}{\partial \zeta}\right) + \left(\frac{\partial w}{\partial \varphi}\right)\left(1 + \frac{\partial w}{\partial \zeta}\right).$$

Using first of equalities (4), we shall eliminate $\rho(\mu, \varphi, \zeta)$ from the equations (6). Thus we receive

$$g_{11} = \left(\frac{\partial R}{\partial \psi}\left(1 + \frac{\partial \theta}{\partial \varphi}\right) + \frac{\partial R}{\partial z}\frac{\partial w}{\partial \varphi}\right)^2 + R^2\left(1 + \frac{\partial \theta}{\partial \varphi}\right)^2 + \left(\frac{\partial w}{\partial \varphi}\right)^2, \qquad (7)$$

$$g_{22} = \left(\frac{\partial R}{\partial \psi}\frac{\partial \theta}{\partial \zeta} + \frac{\partial R}{\partial z}\left(1 + \frac{\partial w}{\partial \zeta}\right)\right)^2 + R^2\left(\frac{\partial \theta}{\partial \zeta}\right)^2 + \left(1 + \frac{\partial w}{\partial \zeta}\right)^2.$$

Rates of elongations for lines of material coordinates are determined by formulas

$$\lambda_1 = \sqrt{\frac{g_{11}}{g_{11}^0}}, \lambda_2 = \sqrt{\frac{g_{22}}{g_{22}^0}} \qquad (8)$$

Variation χ of angle between these lines is determined by equality

$$g_{12} = \sqrt{g_{11}^0 g_{22}^0} \lambda_1 \lambda_2 \sin \chi. \tag{9}$$

Value χ and extensional strains of coordinate lines, i.e. magnitudes

$$\epsilon_1 = \lambda_1 - 1, \epsilon_2 = \lambda_2 - 1, \tag{10}$$

will be considered as macroscopical measures of the shell deformation. On the basis of [1], [3] the densities u of its potential strain energy, related to unit of the area of non-stressed shells, were calculated as functions of χ, ϵ_1, ϵ_2 and μ for various reinforcing textile structures of the shells.

Considering (4) – (10), we can present u as some function Φ depending on displacements $\theta(\mu, \varphi, \zeta)$, $w(\mu, \varphi, \zeta)$ of the shell particles and the first derivations of these displacements:

$$u = \Phi\left(\mu, \varphi, \zeta, \theta, w, \frac{\partial \theta}{\partial \varphi}, \frac{\partial \theta}{\partial \zeta}, \frac{\partial w}{\partial \varphi}, \frac{\partial w}{\partial \zeta}\right). \tag{11}$$

In this work it will be supposed that all points of the bottom edge γ_1 of the soft shell have coordinates $\zeta = 0$, and points of the top edge γ_2 – have coordinates $\zeta = H$. Thus potential energy of the whole shell will be equal

$$U = \int_0^H \int_0^{2\pi} u \sqrt{g_{11}^0 g_{22}^0 - g_{12}^0{}^2} d\varphi d\zeta. \tag{12}$$

It must be remarked that this supposition always can be fulfilled for soft shells. But it can lead to formation of the shell's wrinkles and creases. In this case the numerical processes as rule give no results [4].

We will restrict the paper to investigation of displacements $\theta(\mu, \varphi, \zeta)$ and $w(\mu, \varphi, \zeta)$, satisfying to conditions of the shell's edges fixity. These conditions are given by equations

$$\theta(\mu, \varphi, 0) = \theta_0(\varphi) + \mu\theta_0^*(\varphi), \tag{13}$$
$$\theta(\mu, \varphi, H) = \theta_1(\varphi) + \mu\theta_1^*(\varphi),$$
$$w(\mu, \varphi, 0) = w_0(\varphi) + \mu w_0^*(\varphi),$$
$$w(\mu, \varphi, H) = \theta_1(\varphi) + \mu w_1^*(\varphi),$$

where θ_0, θ_0^*, θ_1, θ_1^*, w_0, w_0^*, w_1 and w_1^* are arbitrary given functions.

The analysis of the deformed state of the shell is based on variational methods under the supposition of elasticity of the shell material; therefore the equilibrium state of the shell reduces its potential energy defined by (12) to a minimum.

We shall assume that the shell is thin linearly elastic membrane reinforced by net with rectangular cells. Moreover for convenience of calculations it will be supposed that the membrane much more compliant then the net is. In this case formula (12) can be written in the form

$$U = \int_0^H \int_0^{2\pi} \left(k_1 \epsilon_1^2 + k_2 \epsilon_2^2 + \gamma F(\epsilon_1, \epsilon_2, \chi) \right) f(\varphi, \zeta) d\varphi d\zeta, \qquad (14)$$

where k_1 and k_2 are the coefficients describing elastic properties of the shell; production γF is density of membrane's strain energy and parameter γ will be considered as small in respect to k_1 and k_2; function f in (14) is defined by

$$f(\varphi, \zeta) = \sqrt{g_{11}^0 g_{22}^0 - g_{12}^0}. \qquad (15)$$

3 Scheme of Finite Elements Method

Minimization of the functional U we shall accomplish by the final elements method.

Definitional domain of functions $\theta(\mu, \varphi, \zeta)$ and $w(\mu, \varphi, \zeta)$ is subjected to triangulation according to the scheme shown in Fig. 3.

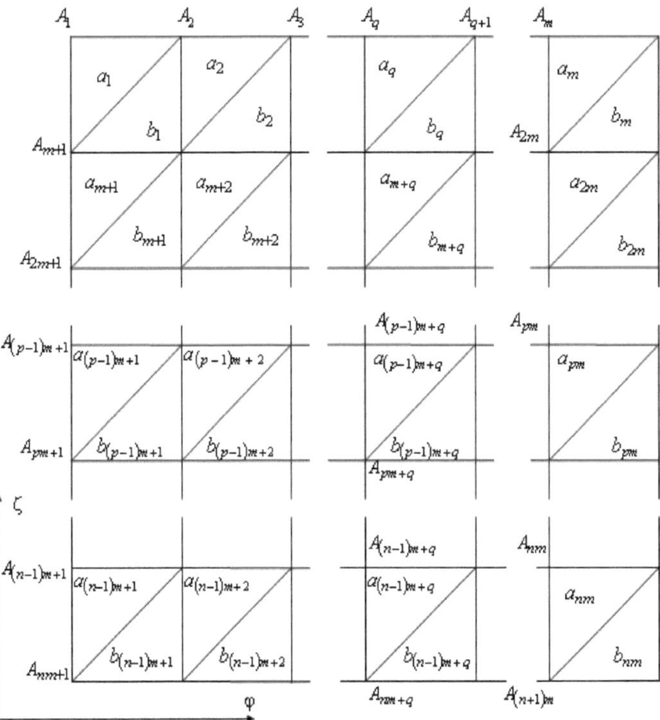

Fig. 3 Triangulation of the function domain

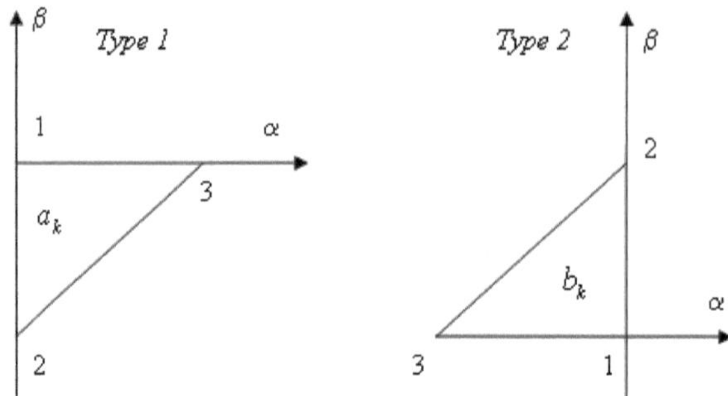

Fig. 4 Two types of finite element

Without essential loss of generality one could suppose legs of all triangles to be equal, so that the following identities hold on

$$s = \frac{2\pi}{m} = \frac{H}{n},\tag{16}$$

where m and n are some integers.

Let us, not counting the rightmost nodes, number the other nodes A_1, A_2, ..., $A_{m(n+1)}$, as it is shown on Fig. 3. At that for every A_k in which p-th string and q-th column of nodes are crossed we have

$$k = (p-1)m + q,\tag{17}$$

As it can be seen in Fig. 3, we deal with two types of elements. As shown on Fig. 4 in finite elements of type 1 and type 2 we will enter local co-ordinates $(\alpha\beta)$. For elements of type 1 and 2 we will introduce three functions

$$f_1 = 1 + \frac{\beta - \alpha}{s}, f_2 = -\frac{\beta}{s}, f_3 = \frac{\alpha}{s},\tag{18}$$

and functions and functions

$$p_1 = 1 + \frac{\alpha - \beta}{s}, p_2 = \frac{\beta}{s}, p_3 = -\frac{\alpha}{s},\tag{19}$$

respectively.

Distinctive feature of these functions is that f_i and p_i equal 1 in apex i and equal zero in other apexes. Everywhere outside of the elements these functions are zero.

The node A_k if $(p > 1 \bigwedge p < n + 1) \bigwedge (q > 1 \bigwedge q < m + 1)$ belongs to domain D_k consisting of elements b_{k-m-1}, a_{k-m}, b_{k-m}, a_k, b_{k-1}, a_{k-1} (see Fig. 3). Taking into account (19) and (20) it is possible to approximate last two formulas of (4) in every of the elements. For example for element b_{k-m-1} we have

$$\psi = s(q-1) + \alpha + T_k p_1 + T_{k-m} p_2 + T_{k-1} p_3, \tag{20}$$
$$z = s(n-p) + \beta + V_k p_1 + V_{k-m} p_2 + V_{k-1} p_3,$$

For other elements approximations are analogous.

It must be kept in mind that all considered functions are periodical in respect to φ and period equals 2π. Because of that the nodes of the leftmost and of the rightmost columns in every string of the nodes have to be identified. The equations (4) in elements including such nodes are approximated by formulas analogous to (21).

Using such approximations for ψ and z calculation of U by formula (14) give us U as function of coefficients T_i and V_i, $i = m+1, m+2, ..., m(n-1)$. By dint of equating the partial derivatives of this function with respect to mentioned coefficients with zero, a system of algebraic equations could be obtained as the solution of which the coefficients could be found.

At the topmost and bottommost strings of nodes in accordance to (13) the values of T_i and V_i may be given arbitrary. Because of that we will introduce the following designations

$$X_{2k-1} = T_{k+m}, X_{2k} = V_{k+m}, k = 1, 2, ..., m(n-1). \tag{21}$$

In this case U transforms into function of X_i, $i = 1, 2, ..., 2m(n-1)$.

4 Equations for Calculation of Coefficients X_k

Conditions of minimality of energy U are represented by equalities

$$\frac{\partial U}{\partial X_k} = 0. \tag{22}$$

After differentiation of integral (15), we shall receive

$$\int\int_{D_k} \left(2k_1\epsilon_1 \frac{\partial \epsilon_1}{\partial X_k} + 2k_2\epsilon_2 \frac{\partial \epsilon_2}{\partial X_k} + \left(\frac{\partial F}{\partial \epsilon_1} \frac{\partial \epsilon_1}{\partial X_k} + \frac{\partial F}{\partial \epsilon_2} \frac{\partial \epsilon_2}{\partial X_k} + \frac{\partial F}{\partial \chi} \frac{\partial \chi}{\partial X_k} \right) \right) f d\varphi d\zeta = 0. \tag{23}$$

In order to analyze this nonlinear equations set let us differentiate all equations with respect to parameter μ. As the result a linear equations set in regard to derivatives $\frac{dX_k}{d\mu}$ is obtained.

The last equation set may be represented in form

$$C(\mu, X(\mu)) \frac{dX}{d\mu} = B(\mu, X(\mu)), \tag{24}$$

where C is a matrix of a format $r \times r$, B is a vector of length r, X is a required vector of the length $r = 2(m-1)(n-2)$.

Thus, the system (26) represents system of the ordinary differential equations from which should be found $T_k(\mu)$ and $V_k(\mu)$.

Solution of this system leads to Cauchy problem if $T_k(0)$ and $V_k(0)$, or more definitely $X_k(0)$, are known values.

Having written down (26) in the form of

$$\frac{dX}{d\mu} = C^{-1}\left(\mu, X(\mu)\right) B\left(\mu, X(\mu)\right),\tag{25}$$

we can construct required solution by the recurrent procedure

$$X(\Delta\mu) = X(0) + \frac{dX}{d\mu}\big|_{\mu=0}\,\Delta\mu = X(0) + C^{-1}\left(0, X(0)\right) B\left(0, X(0)\right)\Delta\mu,$$

$$X(\mu_k + \Delta\mu) = X(\mu_k) + \frac{dX}{d\mu}\big|_{\mu=\mu_k}\,\Delta\mu = X(\mu_k) + C^{-1}\left(\mu_k, X(\mu_k)\right) B\left(\mu_k, X(\mu_k)\right)\Delta\mu.$$

$$\tag{26}$$

Thus, calculation of the shell's deformation at any value μ can be carried out, if the decision of this problem is known at $\mu = 0$, i.e. if is known initial conditions for Cauchy problem related to equation (28).

It is possible to specify various ways of construction such initial conditions. The elementary way consists in considering one-parametrical family of solids with parameter μ. There is an obvious trivial decision $X = 0$, if this family is such that at the some $\mu = \mu_*$ the surface of a body has the demanded form, and at $\mu = 0$ this surface coincides with a surface of not deformed shell.

This way we shall illustrate by dint of following example.

Example. Let's consider problem about the cylindrical shell which is a tube with radius R_0 and length H that is put on the solid's surface given by following equation
$$R(z) = R_0 + \mu z + A\mu\sin\psi,\tag{27}$$

where μ is above mentioned parameter, A is some arbitrary chosen parameter.

In this case the zero value of parameter μ it is possible to confront with such position of the shell on the solid's surface that this shell won't be deformed. This exact solution can be used as beginning condition in investigation of Cauchy problem for calculation of the shell's state. Boundary conditions will be chosen so that one edge of the shell corresponds to coordinate $z = 0$, and another – to coordinate $z = H$.

In this case
$$g^0_{11} = R_0^2, g^0_{12} = g^0_{21} = 0, g^0_{22} = 1.\tag{28}$$

$$g_{11} = \mu^2\left(A\cos\psi\left(1 + \frac{\partial\theta}{\partial\varphi}\right) + \frac{\partial w}{\partial\varphi}\right)^2 + (R_0 + \mu z + A\mu\sin\psi)^2\left(1 + \frac{\partial\theta}{\partial\varphi}\right)^2 + \left(\frac{\partial w}{\partial\varphi}\right)^2,$$

$$g_{22} = \mu^2\left(A\cos\psi\frac{\partial\theta}{\partial\zeta} + \left(1 + \frac{\partial w}{\partial\zeta}\right)\right)^2 + (R_0 + \mu z + A\mu\sin\psi)^2\left(\frac{\partial\theta}{\partial\zeta}\right)^2 + \left(1 + \frac{\partial w}{\partial\zeta}\right)^2.$$

$$\tag{29}$$

5 Definition of Pressure of the Shell on a Surface of the Solid

For definition of pressure of the shell on the surface of the solid the equilibrium equations are considered. In the vector form they look like [1]:

$$\frac{\partial}{\partial \varphi}\left(\vec{\sigma}_1 | \frac{\partial \mathbf{r}}{\partial \zeta}|\right) + \frac{\partial}{\partial \zeta}\left(\vec{\sigma}_2 | \frac{\partial \mathbf{r}}{\partial \varphi}|\right) + \mathbf{q}\sqrt{g_{11}^0 g_{22}^0 - (g_{12}^0)^2} = 0. \qquad (30)$$

Here stresses $overrightarrow\sigma_1$ and $overrightarrow\sigma_2$ are determined by formulas:

$$\vec{\sigma}_1 = \mathbf{e}_1\sigma_{11} + \mathbf{e}_2\sigma_{12}, \qquad (31)$$
$$\vec{\sigma}_2 = \mathbf{e}_1\sigma_{21} + \mathbf{e}_2\sigma_{22},$$

where

$$\mathbf{e}_1 = \frac{\partial r}{\partial \varphi} / | \frac{\partial r}{\partial \varphi} |, \mathbf{e}_2 = \frac{\partial r}{\partial \zeta} / | \frac{\partial r}{\partial \zeta} |, \qquad (32)$$

$$\sigma_{11} = \frac{1}{\lambda_2\sqrt{g_{11}^0 g_{22}^0}}\left(\frac{\partial u}{\partial \lambda_1} + \frac{\cot \chi}{\lambda_1}\frac{\partial u}{\partial \chi}\right), \qquad (33)$$

$$\sigma_{22} = \frac{1}{\lambda_1\sqrt{g_{11}^0 g_{22}^0}}\left(\frac{\partial u}{\partial \lambda_2} + \frac{\cot \chi}{\lambda_2}\frac{\partial u}{\partial \chi}\right),$$

$$\sigma_{12} = \sigma_{21} = -\frac{1}{\lambda_1\lambda_2 \sin \chi\sqrt{g_{11}^0 g_{22}^0}}\frac{\partial u}{\partial \chi}.$$

The value $\mathbf{q}$ is intensity of pressure of the shell on the considered surface. For definition of normal component of $\mathbf{q}$, it is enough to write down the equation (45) in projections on a normal to the solid's surface vector.

6 Conclusions

The numerical finite elements method for investigations of strains and stresses of some reinforced soft shells and also their pressure on surfaces of some enveloped by them solids was developed. The method is applicable for various forms of the solids and various initial forms and constitutive equations of the shells. The method leads to the nonlinear algebraic equations sets for computation of coefficients of some linear approximations of calculated characteristics considered problems. The solutions of these nonlinear equations were researched by methods of prolongation by parameter. Some variable characteristic of the solid form was chosen as the parameter. It may be noted that investigation of these equations is the most difficult part of considered problems. Proposed in present paper method of reducing these equations' solution to Cauchy problem often gives no results because irregularities of numerical procedures required in calculations. Such irregularities may be caused by specific properties of boundary conditions and as well by unevenness of solids or shells surfaces.

References

1. Dyatlova, P., Polyakova, E., Chaikin, V.: Constitutive equations for soft shells with network microstructures. In: Proceedings of XXXIV Summer school Advanced problems in mechanics, Saint-Petersburg, pp. 146–153 (2006)
2. Polyakova E.V., Chaikin V.A.: Applied problems of a mechanics of soft shells and fabrics. Saint-Petersburg, SPSUTD, p. 193 (2006) (Russian)
3. Chaikin, V.A., Polyakova, E.V.: Fundamentals of a mechanics of soft shells and fabrics. Saint-Petersburg, SPSUTD, p. 101 (2004) (Russian)
4. Dyatlova, P., Polyakova, E., Chaikin, V.: On the finite elements method for investigations of the interaction of soft shells with solids' surfaces. In: Proceedings of XXXV Summer school "Advanced problems in mechanics", Saint-Petersburg, pp. 85–92 (2007)

Polina Dyatlova, St-Petersburg State University of Technology and Design, Bolshaya Morskaya St. 18, St.-Petersburg, 191186, Russia

A 3D Stochastic Model of the Cell-Wall Interface during the Rolling

N. Mefti and J.F. Ganghoffer

Abstract. The rolling of biological cells is analysed, from the modeling of the local kinetics of successive attachment and detachment of bonds, occurring at the interface between a single cell and the wall of the endothelium. These kinetics correspond to a succession of creation and rupture of ligand-receptor molecular connections, under the combined effects of mechanical, physical (both specific and non specific) and chemical external interactions. A three-dimensional model of the interfacial molecular rupture and adhesion kinetic events is developed in the present contribution, as an extension of a 2D model with an elastic behaviour of the connections (Mefti et al. 2006). From a mechanical point of view, we assume that the cell-wall interface is composed of two elastic shells, namely the wall and the cell membrane, linked by rheological elements, representing the molecular bonds. Both the time and space fluctuations of several parameters are described by the stochastic field theory. Numerical simulations emphasize the rolling phenomenon, in terms of the time evolution of the number of molecular connections and of the rolling angle.

Keywords: Cell adhesion; ligand-receptor connections; Rolling; Stochastic fields; Rupture; Viscoelasticity; statistical nonlocality.

1 Introduction

Cell adhesion is an important phenomenon in biology, especially in the immune defence, wound healing, and the growth of tissues. It is a multiscale phenomenon involved in *cell rolling* and cell migration (Bongrand et al., 1982; Bongrand and Benoliel, 1999), due to the induced *cell motility*.

Recent works providing computational approaches of cells adhering to a wall under shear flow include those of (Chapman and Cokelet, 1998) who modelled the

N. Mefti and J.F. Ganghoffer
LEMTA ENSEM, 2 A^{nue} Forêt de Haye, BP 160. 54054 Vandoeuvre lès Nancy cedex, France

N. Mefti
LCPC. 58, boulevard Lefebvre, 75732 Paris Cedex 15, France

J.-F. Ganghoffer, F. Pastrone (Eds.): Mech. of Microstru. Solids, LNACM 46, pp. 57–70.
springerlink.com

flow past multiple adherent leukocytes in blood vessels (the adhesion process itself is not considered), (Korn and Schwarz, 2008), whereby states of motion are pictured in a state diagram depending upon the rates of bond formation and rupture, considering the white blood cells as rigid spheres, (Khismatullin and Truskey, 2005) who performed 3D simulations of receptor-mediated adhesion of a spherical viscoelastic leukocyte to a plate, using the spring-peeling kinetic model to describe the receptor-ligand interaction and modelling the leukocyte as a compound viscoelastic drop.

The main objective of this contribution is to model and simulate the local *kinetics of bond rupture and formation*, within a *stochastic framework*, focusing on a single cell approaching the extracellular wall. Compared to the previously mentioned work, a quite simplified description of the fluid flow shall be provided, and the focus shall instead be on the consideration of the kinetics of ligand-receptor pairs adhesion and rupture, under the combination of both specific and non specific forces.

Several works in the literature describe the cell adhesion, either from an experimental or a theoretical point of view. One of the most widely spread experimental technique is the micropipette technique, which is used to determine the adhesion force (Skalak and Evans, 1984; Evans and Needham, 1987; Evans, 1992; Evans and Ritchie, 1997; Zhao et al., 2001). As an alternative, the microscopy technique has been used by (Bruinsma and Sackman 2001) and (Simon, 2002) to describe cell display. As far as the theoretical approaches are concerned, one may distinguish the probabilistic and the deterministic methods. In the probabilistic approaches, some or all parameters of the model are the result of a probabilistic calculus (contrary to the deterministic approaches). In several models, the kinetics of adhesion is described by two parameters, namely the forward and reverse rate constants (Bongrand et al., 1982, Bongrand, 1999). These parameters correspond to the duration and the number of stops in the case of the flow of a cell along a wall characterized by the presence of adhesion molecules. Several models use these parameters in deterministic (Dembo et al., 1998; Bell et al., 1984, Hammer and Lauffenburger, 1987; Ndri et al, 2001, Coombs et al., 2004; Oliver and Jacobson, 1994; Robert et al., 1990; Richert et al., 2004, Turner and Sherrat, 2002) or probabilistic approaches (Cozens et al., 1990; Evans and Ritchie, 1997, Lavalle et al., 1997; Haussy and Ganghoffer, 2001). Further models describe the adhesion of a population of cells (Takano et al., 2003; Mochizuki, 2002; Mochizuki et al., 1996; Wiliams and Bjerknes, 1972).

The kinetics of bond formation and rupture shall herewith be modelled as a stochastic process, but at the scale of individual bonds, and not at the level of the global kinetic level.

2 Sketch of the Cell-Wall Interactions

Rolling of biological cells corresponds to the slowing down of a moving cell (along the wall of an extracellular matrix or a substrate), followed by the capture of the cell by the endothelium. The rolling is the result of complex molecular kinetic events of simultaneous creation and rupture of molecular connections, namely the ligand-receptor bonds, involving specific proteins. These connections are submitted to the action of the fluid flow, the Van der Waals attractive interactions, the electrostatic repulsion (Bongrand et al., 1982) and affinity forces, in combination with their extensibility. The fluid flow (force) around the cell is transmitted from the cell membrane to the contact interface (Mefti et al., 2006). (Bell et al., 1984), Fig 1.

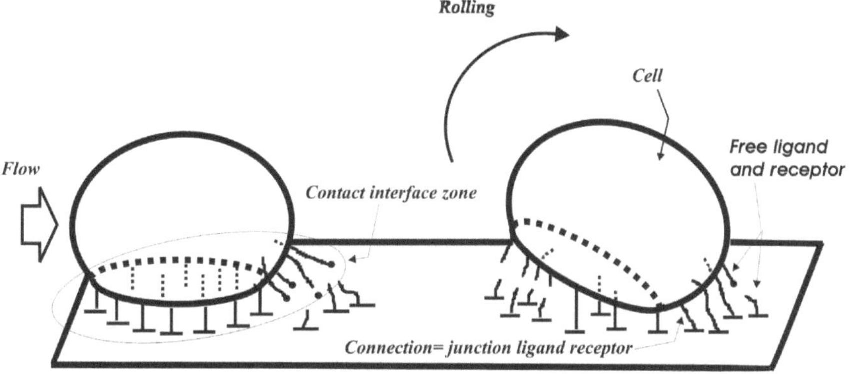

Fig. 1 Rolling of a cell connected to the endothelium by ligand-receptor pairs

Since the specific *signalisation* during the adhesion leads the local increase of the cell membrane stiffness (in the interface zone, fig. 1), consecutive to the change of the cytoskeleton structure (Bongrand et al., 1982), we model the cell membrane (at the contact interface zone, Fig. 1) and the wall as two elastic plates, endowed with high stiffness. The plates are linked by rheological elements, representing the molecular connections (Zhu, 2000). The cell itself is supposed not to deform during the rolling, hence it follows the motion of the interface as a rigid body.

The cells to endothelium interactions are traduced in terms of forces, further involved in the virtual work principle. The external forces acting on the interface are:

-    The fluid force: in the case of blood, the action of plasma is periodic and generates a periodic force $F_{fluid}(t)$ characterized by two adjustable parameters, namely its intensity F_0 and inclination, parameterized by the two angles $\hat{\beta}_{rand}(t), \hat{\theta}_{rand}(t)$ (Fig. 3). We herewith do not solve the Navier-Stokes equations, but consider that the cell does not perturb the imposed fluid flow;

-    The Van der Waals forces (attractive), are obtained by the derivation of the expression of the Van der Waals energy between two elastic plates (Tadmor, 2001), as

$$F_{attr} = \frac{A_0\, l_{a_0}}{6\pi}\left(\frac{1}{(S+D_0)^3} - \frac{1}{S^3}\right) \tag{1.1}$$

with A_0 the Hamacker constant, l_{a_0} the length of the interaction zone, S the cell-wall distance, and D_0 the thickness of the cell membrane;

-        The electrostatic force – of a repulsive nature - (Bongrand et al., 1982), resulting from the presence of negative charges on the cell membrane. This force is obtained by the minimization of the phenomenological energy (Bell et al., 1984):

$$F_{rep} = -\frac{\chi}{S}e^{-S/\tau}\left(\frac{1}{S}+\frac{1}{\tau}\right) \tag{1.2}$$

with χ a compressive parameter describing the ease with which the connections may be compressed, and τ a thickness parameter (the mean length of the ligand).

The main *non specific interactions* considered are the Van der Waals forces and the electrostatic forces; other forces, such as the solvent influence, the influence of the surrounding ions and hydrogen bonds (Bongrand, 1982), are here discarded.

2.1 *Failure of the Cell-Ligand Connections*

Several mechanisms are responsible of the failure of the molecular connections, being either active or passive (Bongrand et al., 1999). Rupture is assumed to occur only by a pulling effect (mode I): thus, the ith connection fails if the traction force it sustains exceeds a threshold value, viz

$$F\left(x_i, y_i, t\right) \geq \tilde{F}_{rupt}^{rand}\left(x_i, y_i\right) \tag{2.1}$$

$F(x_i, y_i, t)$ therein is the total external force acting on the ith connection having the (discrete) position $\left(x_i, y_i\right)$, while $\tilde{F}_{rupt}^{rand}(x_i, y_i)$ is the limit of failure of the same connection.

It is natural to consider that the intensity of the forces of the fluid applied to the connections decreases in the direction of the flow: hence, we here assume a linear decrease of the fluid forces. The spatial distribution of forces involves the notion of rows: the support of each force (point of application) is identified by its position in the corresponding transversal row. The junctions between the connections and the cell membrane correspond to the nodes. The following node indexing is used: a given node is identified by two parameters (i, j), which correspond respectively to the position of the transversal and the longitudinal row. Accordingly, the connection forces are written $F_i^{(j)}(t)$. Using the equilibrium equations and the calculated shape of the rupture force distribution gives

$$F_i^{(1)}(t) = \frac{F_{fluid}(t)}{n^{(1)} + \sum_{j=1}^{NR-1} n^{(j)}\left(\frac{l_{j+1}}{l_j}\right)} \tag{2.2}$$

Furthermore, the forces acting on the successive longitudinal rows are expressed by:

$$F_i^{(2)} = \left(\frac{l_2}{l_1}\right) F_i^{(1)}; \quad F_i^{(3)} = \left(\frac{l_3}{l_1}\right) F_i^{(1)}; \quad F_i^{(NR-1)} = \left(\frac{l_{NR-1}}{l_1}\right) F_i^{(1)} \qquad (2.3)$$

NR, l_i and $n^{(i)}$ are respectively the number of longitudinal rows, the distance between the extremity and node i, and the number of bonds along the transversal row i. The set of equations (2.2), (2.3) gives the forces applied to the molecular connection under the effect of the net external force $F(t)$ evaluated as:

$$F(t) = F_{fluid}(t) + F_{rep} - F_{attr} \qquad (2.4)$$

Since the established connections are the result of the junction between adhesion molecules (receptors on the cell and ligands on the endothelium) occurring during the rolling, one may assume that these junctions are established under non stationary conditions of the surrounding fluid (temperature, pressure…) (Mefti et al., 2006). Consequently, these variations may lead to spatial fluctuations of the connections properties. In order to simplify the model, we assume that the fluctuation in the limit of failure is characterized by a unique parameter, which is modelled within the *theory of stochastic fields*. The spatial distribution of the limit of failure is then given by

$$\tilde{F}_{rupt}^{rand}(x_i, y_i) = F_0 \left(1 + \delta_p \frac{f_{rand}(x_i, y_i)}{f_{max}} \right) \qquad (2.5)$$

The amplitude F_0 is the limit of failure of a connection in the case of a uniform distribution (it can be measured), δ_p the value of the maximal fluctuation of the Gaussian stochastic field $f_{rand}(x_i, y_i)$ in (2.5), and f_{max} its maximal value. The *Gaussian stochastic field* (here describing a spatial fluctuation) is obtained by the *spectral approach* of (Shinozuka and Deotadis, 1991; Shinozuka et al., 1999):

$$f_{rand}(x_i y_i) = \sqrt{2} \, Re\{ \sum_{m=0}^{M_1-1} \sum_{n=0}^{M_2-1} \sqrt{4S_{00}(k_{1m}, k_{2n})} \times \sqrt{\Delta k_1 \Delta k_2} \, e^{i\phi_{mn}} e^{i(k_{1m}x + k_{2n}y)} \}$$

$$(2.6)$$

with the discrete spatial positions of the bonds:

$$x_i = p\Delta x \quad p = 1, 2, 3 \ldots\ldots M_1; \quad y_i = q\Delta y \quad q = 1, 2, 3 \ldots\ldots M_2$$

The function $S_{00}(k_{1m}, k_{2n})$ in (2.6) represents the *Spectral Density of Power* (SDP), accounting for the long range correlations between the values of the failure limit; the SDP is the Fourier transform of the self-correlation function (according to the Bochner-Wienwer-Khintchine theorem for stationary processes). We presently use the SDP given in (Nour et al., 2003):

$$S_{00}(k_1,k_2)=\sigma^2\,\frac{b_1 b_2}{4\pi}\,e^{\left[-\left(\frac{b_1 k_1}{2}\right)^2-\left(\frac{b_2 k_2}{2}\right)^2\right]} \tag{2.7}$$

The *correlation lengths* (b_1,b_2) determine the extent of the spatial range of interactions in the directions x and y respectively, and σ is the standard deviation (Nour et al., 2003). The spatial discretization steps $(\Delta x,\Delta y)$ depend on the periodicity lengths (L_x,L_y) (Shinozuka et al., 1999) according to:

$$\Delta x = L_x / M_1; \quad \Delta y = L_y / M_2 \tag{2.8}$$

(M_1,M_2) are the number of harmonics in the directions x and y respectively (the number of spatial discretization points), (k_{1m},k_{2n}) the wave numbers in the directions x, y respectively and $(\Delta k_{1m},\Delta k_{2n})$ the discretization steps of the wave numbers, determined from two integers N_1, N_2 as $\Delta k_{1m} = \dfrac{k_{1u}}{N_1}; \quad \Delta k_{2n} = \dfrac{k_{2u}}{N_2}$. The parameters k_{1u}, k_{2u} are respectively the upper limits of the wave numbers in the directions x, y, such that

$$-k_{1u} \le k_{1m} \le +k_{1u}; -k_{2u} \le k_{2n} \le +k_{2u}$$

When the rupture of the connection (i,j) occurs, the applied force $F_i^j(t)$ is transferred to the other connections located in a circular influence zone of radius R_D, resulting in a force jump ΔF_k^l at the nodal position (i,k). Connections close enough to the connection (i,j) recover a proportion of the initial force $F_i^j(t)$ being redistributed. Introducing $L_{j,j}^{k,l}$ as the distance between the connections (i,j) and (k,l), and assuming that the force transfer is linear and decreasing (Fig. 2), the equilibrium equations give the force jump

$$\Delta F_{i+1}^j(t+\Delta t)=\frac{F_i^j(t)}{4+(R_D-L_{i,i+1}^{j,j})}\frac{1}{(4R_D-4L_{i,i+1}^{j,j}-L_{i+1,i-1}^{j,j-1}-L_{i+1,i-1}^{j,j+1}-L_{i+1,i+1}^{j,j-1}-L_{i+1,i+1}^{j,j+1}}$$

$$\tag{2.9}$$

denoting $L_{i,k}^{j,l}$ the distance between the pair of connections (i,j) and (k,l).

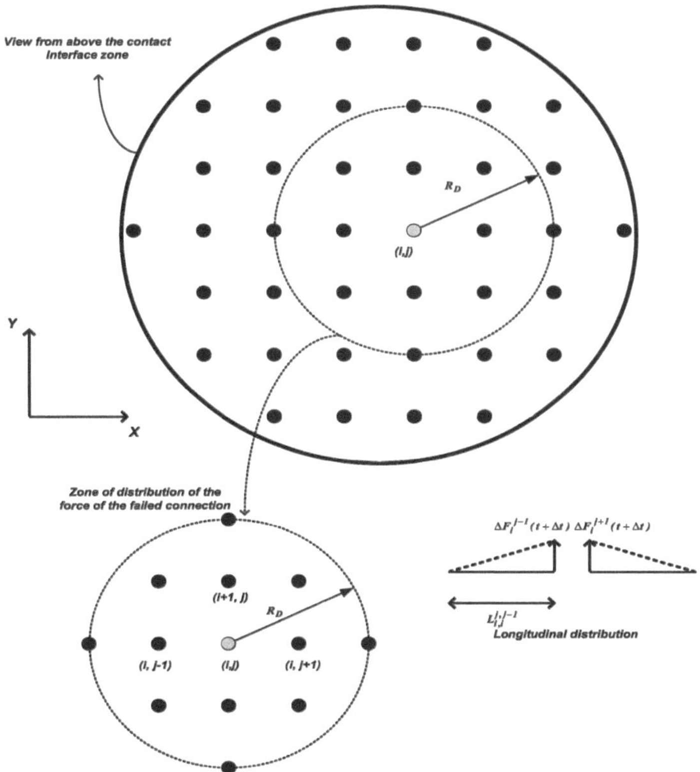

Fig. 2 Rupture of the (i,j) connection and transfer of the force to the remaining connections

2.2 *Adhesion of Molecular Connections*

The generation of molecular bonds is the result of the close contact of complementary adhesion molecules (ligands and receptors) on a portion of the cell outer surface, (Bongrand and Benoliel, 1999), Fig. 3. We assume that the ligand is subjected to the action of the thermal agitation and to the molecular specific interactions (Fig. 3), namely Van der Waals forces, electrostatic interactions, solvent influence, influence of the surrounding ions and hydrogen bonds (Bongrand et al., 1999).

The Brownian force is a quasi random phenomenon, according to the physical picture of forces resulting from the shocks between the fluid particles and the ligands. We model this action relatively to each ligand by a periodic force, characterized by both a random orientation and intensity (Mefti et al., 2006). In order to simplify the problem, the selected specific interactions being considered are restricted to Van der Waals forces. Consequently, the intensities of the external forces applied to the ith ligand are:

$$F_{aff_j} = \frac{4}{k_T T}\left(\frac{\mu_a \mu_b}{4\pi\varepsilon_0}\right)\frac{1}{r^7} \; ; \;\; F_{brow_j} = F_{0_j}\sin(\omega_{brow}t) \qquad (2.10)$$

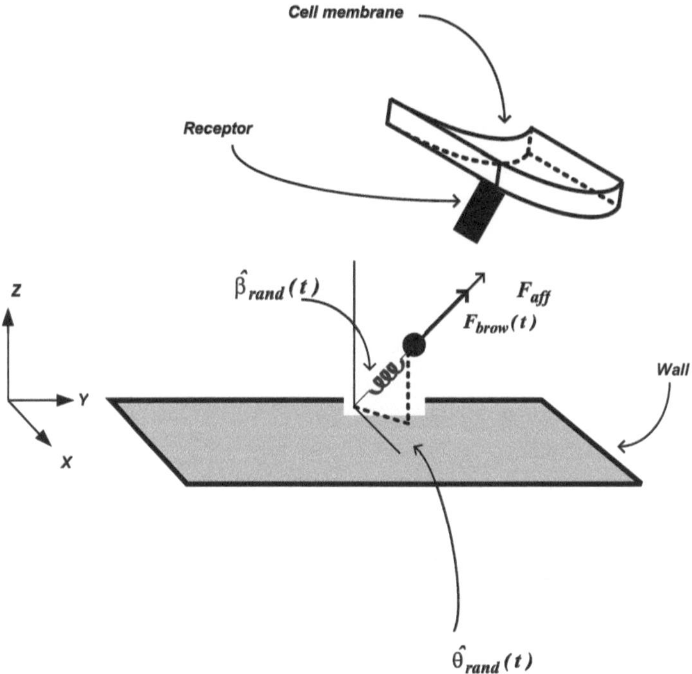

Fig. 3 Forces acting on free adhesion molecules with the ith ligand-receptor (Marques, 2001)

The parameters μ_b, μ_b therein are the dipolar moments of the ligands and receptors respectively, which can be measured (Bamba and N'guessan, 2003) for the adhesion molecules (LFA1...); k_T is the Boltzmann constant, T the temperature, ε_0 the dielectric constant; F_{0_j} is the maximal value of the Brownian force. The considered random parameters are the orientation angles of the ligand $\hat{\beta}_{rand}(t), \hat{\theta}_{rand}(t)$ (Fig. 3) and the pulsation of the solicitation ω_{brow}, both represented in normalized form as:

$$\hat{\beta}_{rand}(t) = \hat{\beta}_{max} \frac{f_\beta^{rand}(t)}{f_\beta^{max}}; \quad \hat{\theta}_{rand}(t) = \hat{\theta}_{max} \frac{f_\theta^{rand}(t)}{f_\theta^{max}}; \quad \hat{\omega}_{brow}(t) = \hat{\omega}_{max} \frac{f_\omega^{rand}(t)}{f_\omega^{max}}$$

$$(2.11)$$

The quantities $f_\beta^{rand}, f_\theta^{rand}, f_\omega^{rand}$ therein are Gaussian stochastic processes (time dependent processes), given by the following series expression:

$$f^{rand}(t) = Re\left\{ \sum_{n=0}^{M-1} B_n \, exp\left[i\left(n\,\Delta\omega\right)\left(p\,\Delta t\right)\right]\right\}$$

$$(2.12)$$

with the amplitudes $B_n = \sqrt{2}A_n e^{i\phi_n}$; $A_n = (2S_{f_0 f_0} (n\Delta\omega)\Delta\omega)^{1/2}$, and the time step given by $t = p\,\Delta t$; $p = 0, 1, 2, \ldots, M-1$ (M is the number of spectral discretization points)

The function $S_{f_0 f_0}$ is the SDP of this Gaussian process, given by (fig. 4)

$$S_{f_0 f_0} = \frac{1}{4}\sigma^2 b^3 \omega^2 e^{-b|\omega|} \qquad (2.13)$$

σ is the standard deviation, b the correlation time, and ω the pulsation. The frequential increment $\Delta\omega$ is given by: $\Delta\omega = {}^{\omega_u}\!/\!_N$, with the condition $N \leq \frac{1}{2}M$, in order to avoid aliasing.

Fig. 4 shape of the SDP with $b = 1s$

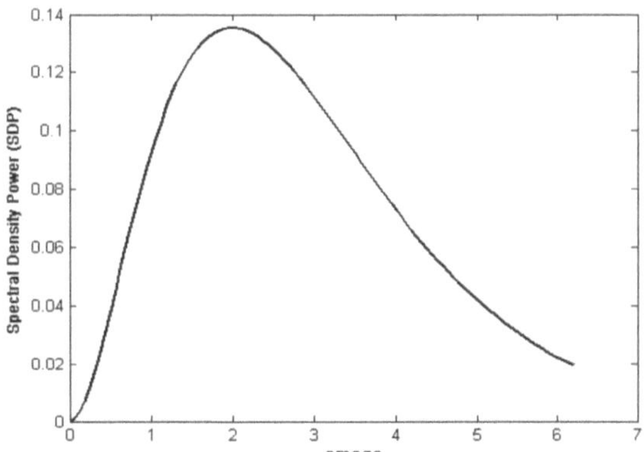

(Shinozuka and Deodatis, 1996). The parameter ω_u corresponds to the upper cut-off frequency. Furthermore, M has to be multiple of 2, while Δt shall satisfy the Heisenberg-Gabor inequality (Shinozuka, 1999).

$$\Delta t \leq \frac{2\pi}{\omega_u}$$

The possible temperature dependence of the stochastic fields has been presently neglected.

The junction between a ligand and a receptor (setting up of bond) is described by a kinematic criterion of minimal proximal distance (Mefti et al., 2006). The motion of the receptors is the result of the rolling, itself being induced by the sequence of rupture / adhesion events of the existing connections (Fig. 2). The dynamical equations of equilibrium of the ith ligand are solved with a finite difference scheme. The damping of molecular connections represents the internal friction occurring at a microscopic scale; the Rayleigh method is presently used for the determination of the damping coefficient, in terms of the mass and rigidity, viz $c_i = \alpha_d + \beta_d m_i$, with α_d, β_d coefficients introduced to correct the damping coefficient (the value of those coefficients is determined from a modal analysis of the structure, (Imbert, 1991)).

Considering a ligand endowed with a viscoelastic behaviour, its motion is expressed by the dynamical equations, wherein the right-hand side of the system of equalities expresses the projection of the stochastic forces responsible for the affinity and Brownian motion:

$$
\begin{cases}
k_i U_j(t) + c_i \dfrac{dU_j(t)}{dt} + m_i \dfrac{d^2 U_j(t)}{dt^2} = (F_{brow_j} + F_{aff_j})\, \sin\!\left(\hat{\beta}_{rand}(t)\right) \cos\!\left(\hat{\theta}_{rand}(t)\right) \\[2ex]
k_i V_j(t) + c_i \dfrac{dV_j(t)}{dt} + m_i \dfrac{d^2 V_j(t)}{dt^2} = (F_{brow_j} + F_{aff_j})\, \sin\!\left(\hat{\beta}_{rand}(t)\right) \sin\!\left(\hat{\theta}_{rand}(t)\right) \\[2ex]
k_i W_j(t) + c_i \dfrac{dW_j(t)}{dt} + m_i \dfrac{d^2 W_j(t)}{dt^2} = (F_{brow_j} + F_{aff_j})\, \cos\!\left(\hat{\beta}_{rand}(t)\right)
\end{cases}
$$

The parameters k_i, m_i, c_i therein are respectively the stiffness, mass and damping coefficient of the ith ligand.

3 Simulation Results

The numerical simulations give the time evolution and the localisation of the rupture / adhesion of molecular bonds, as opposed to global postulated kinetic models. Considering leukocytes, we simulate the behaviour of an interface composed of 60 initially existing connections; the rupture of the bonds depends of the spatial distribution of their limit of failure, according to the chosen SDP function. We consider an interface zone having a circular shape (its radius is $12\mu m$).The parameters have the following values:

$$
l_{a_0} = 3\mu m;\ D_0 = 7nm\,;\ \tau = 10nm;\ \chi = 0.01nN \quad [\text{Bell, 1984}]
$$

$$
m_i = m = 10^{-6} nkg;\ k_i = k = 0.001nN\,/\,nm;\ c_i = c = 0.03nN\,/\,nm
$$
$$
F_0 = 0.07nN;\ \delta_p = 14\%; M_1 = M_2 = 64; N_1 = N_2 = 16; \sigma = 1; b_1 = b_2 = 100nm;\ R_D = 3\mu m
$$
$$
\omega = 4.71rd\,/\,s;\ k_{1u} = k_{2u} = 0.07\ rd/nm;\ \beta_{max} = 0.8rd;\ \theta_{max} = 6.28rd
$$

The interface is further subjected to a fluid force having a resultant of $0.1nN$. The global simulated rupture force is $0.07nN$; this value is in good agreement with the range of measured adhesion forces [1.7pN, 6.7nN], (Bongrand and Benoliel, 1999). The simulations show that the adhesive zone fails in an avalanche manner after $1s$ (Fig. 5).

Initially existing bonds would remain intact for a weaker fluid perturbation, which would prohibit the rolling (all connections need to be broken for rolling to occur).

Fig. 5 Time evolution of the number of connections under rupture. Fluid resultant $0.1nN$

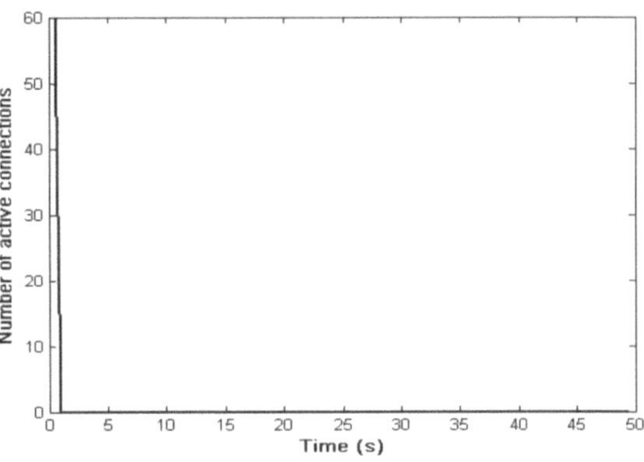

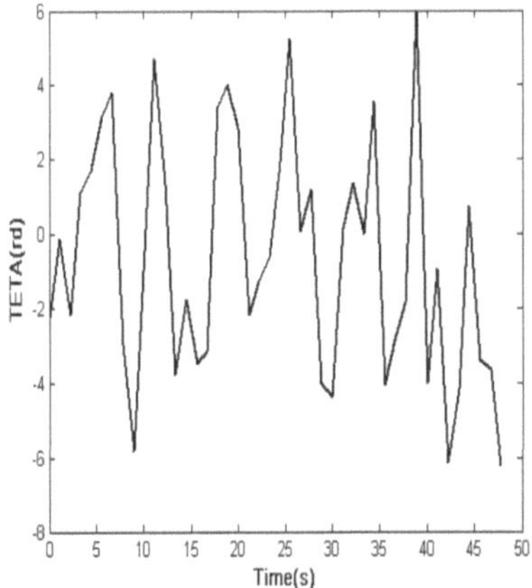

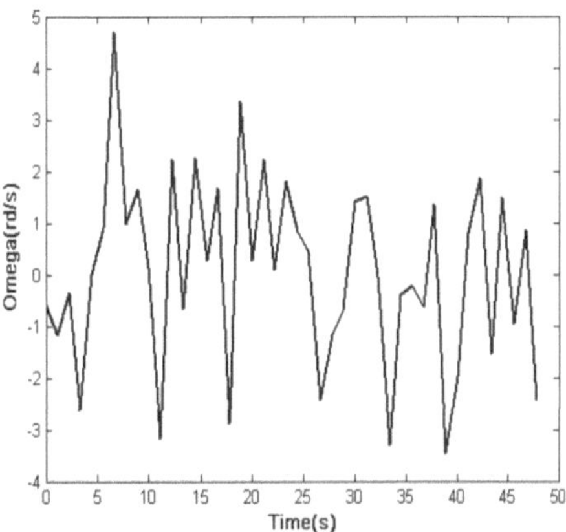

Fig. 6 Time fluctuation of $\hat{\beta}_{rand}, \hat{\omega}_{brow}$ for $M = 50, N = 20, \sigma = 1, b = 1s$

The adhesion of connections is next modelled, considering 25 potential pairs of free adhesion molecules (ligands-receptors pairs). The time evolution of the stochastic parameters describing the orientation of the first ligand is obtained from the equations (2.11), (2.12), (2.13), Fig. 6.

The coupling between the thermal agitation and the specific interactions characterized by a Van der Waals force expressed in (2.10) with $\mu_a = \mu_b = 1debye$ leads to the fast junction of *15* couples of molecules (Fig. 7 a,b), in both cases of a damped and undamped ligand-receptor junctions.

The adhesion kinetics includes two steps: the transitive step corresponds to a period without adhesion (*5s* for the damped system, insert a and *2s* for the undamped system, insert b), followed by the junction between the free adhesion molecules (after respectively *5s* and *2s* for the damped and undamped system). The damping delays the onset of adhesion due to a lower vibration amplitude of the ligands, but does not modify the evolution of the kinetics of adhesion thereafter, exhibiting an avalanche nature, similar to the rupture kinetics.

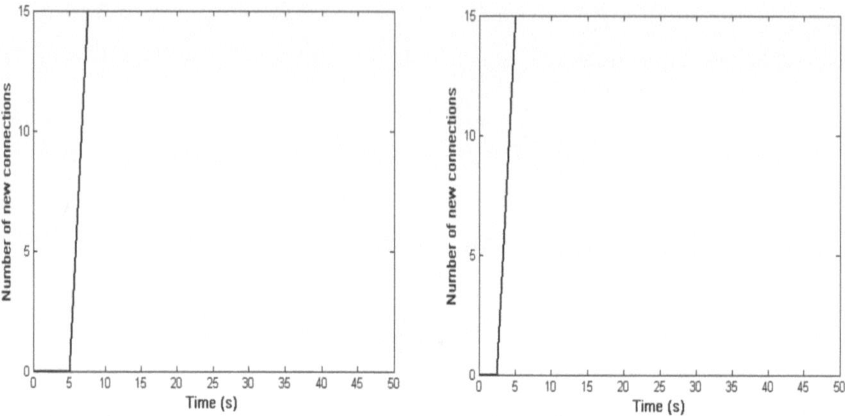

Fig. 7 Time evolution of the creation of new connections with (a) and without damping (b)

4 Perspectives

As a development of those models of the kinetics of attachment / detachment of bonds at the interface between the cell and the endothelium, both the nonlinear elasticity of the bonds (according to statistical models of chain deformation, considering both entropic and energetic contributions) and the deformations of an adhering cell during the cell migration shall be considered, in order to model the cell motility. A combined theoretical-experimental methodology is further necessary in order to both identify the parameters of the model (bond elasticity; specific forces; Brownian forces) and to develop a specific SDP for the biological phenomena, by the use of the FFT approach.

References

Bamba, E., N'guessan, Y.: Nouvelle approche de détermination du moment dipolaire en solution dans des solvents polaires. Revue Ivoirienne des Sciences et Technologies 4, 25–33 (2003)

Bongrand, P., Benoliel, A.M.: Adhésion cellulaire. RSTD 44, 167–178 (1999)

Bongrand, P., Capo, C., Depied, R.: Physics of cell adhesion. Progress in surface Science 12, 217–286 (1982)

Bongrand, P.: Ligand-receptor interactions. Rep. Prog. Phys. 62, 921–968 (1982)

Bell, G.I., Dembo, M., Bongrand, P.: Cell adhesion: competition between non-specific repulsion and specific bonding. Biophys. J. 45, 1051–1064 (1984)

Bruinsma, R., Sackmann, E.: Bioadhesion and the dewetting transition. C. R. Acad. Sci. Paris, 2, Série IV, 803–815 (2001)

Chapman, G.B., Cokelet, G.R.: Flow resistance and drag forces due to multiple adherent leukocytes in postcapillary vessels. Biophys. J. 74, 3292–3301 (1998)

Coombs, D., Dembo, M., Wosfy, C., Goldstein, B.: Equilibrium thermodynamics of cell-cell adhesion mediated by multiple ligand-receptor pairs. Biophysical Journal 86, 1408–1423 (2004)

Cozens, C., Lauffenburger, D., Quinn, J.A.: Receptor-mediated cell attachment and detachment kinetics-probabilistc model and analysis. Biophys. J. 58, 841–856 (1990)

Dembo, M., Torney, D.C., Saxman, K., Hammer, D.: The reaction-limited kinetics of membrane-to-surface adhesion and detachment. Proc. R. Soc. Lond. B 234, 55–83 (1998)

Evans, E.: Equilibrium "wetting" of surfaces by membrane-covered vesicles. Advanced in colloid and Interface Science 39, 103–128 (1992)

Evans, E., Needham, D.: Physical properties of surfactant bilayer membranes: thermal transition- elasticity- rigidity- cohesion and colloidal interactions. J. Phys. Chem. 91, 4219–4228 (1987)

Evans, E., Ricthie, K.: Dynamic strength of molecular adhesion bonds. Biophysical Journal 72, 1541–1555 (1997)

Fenton, G.A.: Simulation and analysis of random fields, PhD thesis, Princeton university (1990)

Hammer, D.A., Lauffenburger, D.A.: A dynamical model for receptor-mediated cell adhesion to surfaces. Biophys. J. 52, 475–487 (1987)

Haussy, B., Ganghoffer, J.F.: Modélisation mécanique du contact globule rouge/vaisseau sanguin: vers la mise en évidence du phénomène de rolling, 16èmecongrès Français de mécanique, 1–6 (2001)

Imbert, J.F.: Analyse des structures par éléments finis. Cepadues Edition (1991)

Khismatullin, D.B., Truskey, G.A.: Three-dimensional numerical simulation of receptor-mediated leukocyte adhesion to surfaces: Effects of cell deformability and viscoelasticity. Phys. Fluids 17, 031505 (2005)

Korn, C.B., Schwarz, U.S.: Dynamic states of cells adhering in shear flow: from slipping to rolling. Phys. Rev. E 77, 41904 (2008)

Lavalle, P., Stoltz, J.F., Senger, B., Voegel, J.C., Schaaf, P.: Red blood cell adhesion on a solid/liquid interface. Poc. Nat. Acad. Scie. USA 93, 15136–15140 (1997)

Lee, H.S., Gheysel, E., Bell, W.R.: Seasonal time serie and autocorrelation function estimation. Série scientifique CIANO 97s-35, Montréal (October 1997)

Mefti, N., Haussy, B., Ganghoffer, J.F.: Mod\'{e}lisation du rolling de globules rouges, 17\`{e}me Congr\`{e}s Fran\c{c}ais de M\'{e}canique, 159 (2005)

Mefti, N., Haussy, B., Ganghoffer, J.F.: Mechanical modeling of the rolling phenomenon at the cell scale, Euromech Colloquium 463, Groningen, Netherlands (2005)

Mefti, N., Haussy, B., Ganghoffer, J.F.: Mechanical modelling of the rolling phenomenon at the cell scale. International Journal of Solid and Structure (2006)

Mochizuki, A.: Pattern formation of the cone mosaic in the Zebrafish Retina: A cell rearrangementmodel. J. theo. Biolo. 215, 345–361 (2002)

Mochizuki, A., Ywasha, Y., Takeda, Y.: A stochastic model for cell sorting and measuring cell-cell adhesion. J. theo. Biolo. 179, 129–146 (1996)

Nri, N., Udaykumar, H.S., Shyy, W., Tay, R.T.S.: Computational modeling of cell adhesion and movement using continuum-kinetic approach. In: BED, Bioengineering Conference ASME, vol. 50, pp. 367–368 (2001)

Oliver, T., Lee, J., Jacobson, K.: Forces exerted by locomoting cells. seminaire in cell biology 5, 139–147 (1994)

Roberts, C.C., Lauffenburger, D., Quinn, J.A.: Receptor mediated cell attachment and detachment kinetics I: probabilistic model and analysis. Biophys. J. 58, 841–856 (1990)

Richert, L., Engler, A.J., Discher, D.E., Picart, C.: Surface measurement of the elasticity of native and cross-linked polyelectrolyte multilayer film, XXIX Congrès de la Société de Biomécanique, Créteil, France (2004)

Shinozuka, M., Deotadis, G.: Simulation of stochastic process by spectral representation. Appl. Mech. Rev. 44(4), 191–203 (1991)

Shinozuka, M.: Monte Carlo solution of structural dynamics. Computers and Structures 2, 855–874 (1972)

Shinozuka, M.: Simulation of multivariate and multidimensional random process. The Journal of Acoustical society of America 49(1), 357–367 (1971)

Shinozuka, M., Lenoe, E.: A probabilistic model for spatial distribution of material properties. Engineering Fracture Mechanics 8, 217–227 (1976)

Shinozuka, M.: Digital simulation of random processes and its applications. Jounal of Sound and Vibration 25(1), 111–128 (1972)

Shinozuka, M., Deodatis, G., Zhang, R., Papageoriou, A.R.: Modeling, synthesis and engineering application of strong earthquake wave motion. Soil Dynamics and Earthquake Engineering 18, 209–228 (1999)

Simon, A.: Intéret de la microscopie de force atomique sur la biofonctionnalisaton de matériaux: caractérisation du greffage et de l'adhésion cellulaire, PhD thesis university of Bordeaux I (2002)

Skalak, R., Evans, E.A.: Mechanics and thermodynamics of biomembranes. CRC Press Inc., Boca Raton (1984)

Tadmor, R.: The London – Van der Waals interactions between objects of various geometries. J. Phys: condens Matter 13, L195–L202 (2001)

Takano, R., Mochizuki, A., Iwasa, Y.: Possibility of tissue separation caused by cell adhesion. J. Theor. Biol. 221, 459–474 (2003)

Turner, S., Sherrat, J.A.: Intercellular adhesion and cancer invasion: a discrete simulation using the extended potts model. J. theo. Biolo. 216, 85–100 (2002)

Williams, T., Bjerknes, R.: Stochastic model for abnormal clone spread through epithelial basal layer. Nature 236, 19–21 (1972)

Zhao, H., Stoltz, J.F., Zhuang, F., Wang, X.: Etude dynamique de l'interaction entre mol\'{e}cules d'adh\'{e}sion \`{a} la surface cellulaire, 15\`{e}me Congr\`{e}s Fran\c{c}ais de M\'{e}canique, Nancy (2001)

Plane Waves in Linear Thermoelastic Porous Solids

Vincenzo Giacobbe and Pasquale Giovine

Abstract. We consider the behavior of porous materials with large irregular holes in the context of a linearized thermoelastic theory which includes inelastic surface effects associated with changes in the deformation of the holes in the vicinity of void boundaries and which generalizes classical voids theories. After we study the dispersion relation for small-amplitude acoustic waves and obtain eight basic waves: two shear optical micro–elastic waves, two coupled transverse elastic waves and four coupled longitudinal thermo-elastic waves.

1   Introduction

For many years the volume fraction of the pores was the only additional degree of kinematical freedom introduced in the voids theory for the study of thermoelastic materials with spherical lacunae finely dispersed in the matrix [2, 13]. But when the holes are large, this scalar microstructure is not sufficient to describe the microdeformation; in fact, the voids theory does not predict size effects in torsion of bars in an isotropic material, while they occur both in bending and in torsion, as observed for bones and polymer foam materials in [11]. A better refinement of the Cauchy theory is necessary to characterize the more complex structure, even if some problem of physical concreteness or of mathematical hardness could arise [3, 9].

Cowin itself [4] pointed out the importance of the shape of the pores in the description of bone canaliculi or of lacunae containing osteocytes: in the human bone, *e.g.*, the lacunae are roughly ellipsoidal with mean values along the axes of about 9 μm, 22 μm and 4 μm. In [11, 14] it was suggested to use the Cosserat theory, but one of us in [6, 7] preferred to generalize the idea of continua with voids (that is, in some sense, complementary to it) by

Vincenzo Giacobbe and Pasquale Giovine
Dipartimento di Meccanica e Materiali, Università degli Studi "Mediterranea",
Via Graziella 1, Località Feo di Vito, 89122 Reggio Calabria, Italy
e-mail: `Giovine@unirc.it`

J.-F. Ganghoffer, F. Pastrone (Eds.): Mech. of Microstru. Solids, LNACM 46, pp. 71–80.
springerlink.com

considering a continuum with an ellipsoidal microstructure which describes bodies whose material elements contain a big pore filled by an inviscid fluid, or an elastic inclusion, both of negligible mass, which could have a microstretch different from and independent of the local affine deformation ensuing from the macromotion and so could allow distinct microstrains along the principal axes of microdeformation, in absence of microrotations (*e.g.*, composite materials reinforced with chopped elastic fibers, porous media with elastic granular inclusions, real ceramics, *etc.*). The voids theory is contained in the present model, when the microstrain is constrained to be spherical.

Here we extend the linear theory [7] of porous solids to the thermoelastic case and include a rate effect in the pores' response, which results in internal dissipation from experimental evidence [10]. After we analyze the propagation of harmonic plane waves and obtain the secular equations governing the two shear optical micro–elastic waves, the two coupled transverse elastic waves and the four coupled longitudinal thermo-elastic waves. The exact or approximate values of the phase velocity, specific loss, attenuation factor and amplitude ratio are discussed for large and small frequencies.

2   Summary of the Model

The porous solid $\mathcal{B}$ is depicted as a continuum with ellipsoidal microstructure and is identified with a fixed region of the three dimensional Euclidean space $\mathcal{E}$, the "natural" reference placement $\mathcal{B}_*$, that is homogeneous and free of residual stresses (see [6, 7]). A generic material element of $\mathcal{B}_*$ is denoted by $\mathbf{x}_*$; thus the thermomechanical behaviour of $\mathcal{B}$, as a continuum with big pores, is described by three smooth mappings on $\mathcal{B}_* \times \Re$ ($\Re$ being the set of real numbers): the spatial position $\mathbf{x} \in \mathcal{E}$ at time τ of the material point which occupied the position $\mathbf{x}_*$ in $\mathcal{B}_*$; the symmetric tensor field with positive determinant $\mathbf{U}$ that describes the changes in the pore structure, namely a left micro Cauchy–Green tensor of deformation; the positive temperature θ.

The spatial position $\mathbf{x}(\mathbf{x}_*, \tau)$ is a bijection, for each τ, between the reference placement $\mathcal{B}_*$ and the current placement $\mathcal{B}_\tau = \mathbf{x}(\mathcal{B}_*, \tau)$ of the body $\mathcal{B}$ and, therefore, the deformation gradient $\mathbf{F} := \nabla \mathbf{x}(\mathbf{x}_*, \tau)$ is a second order tensor with positive determinant. Through the inverse mapping $\mathbf{x}_*(\mathbf{x}, \tau)$ of $\mathbf{x}$, we could consider all the relevant fields in the theory as defined over the current placement $\mathcal{B}_\tau = \mathbf{x}(\mathcal{B}_*, \tau)$ as well as over the the reference placement $\mathcal{B}_*$ of the body $\mathcal{B}$.

The kinetic energy density per unit mass is here the sum of two terms the classical translational one $\frac{1}{2}\dot{\mathbf{x}}^2$ and a term due to the microstructure, $\frac{1}{2}\kappa(\mathbf{U})\dot{\mathbf{U}} \cdot \dot{\mathbf{U}}$, which expresses the inertia related to the admissible expansional micromotions of the pores' boundaries (see, also, [1]); the superposed dot ($\dot{\cdot}$) denotes the material time derivative.

The general system of balance equations which rules the admissible thermo-kinetic processes for porous solids with large irregular voids was proposed in [6]; they are the mass conservation, the Cauchy equation, the micromomentum and moment of momentum balances, the energy balance and the entropy inequality in the Eulerian description:

$$\rho \det \mathbf{F} = \rho_*, \quad \rho \ddot{\mathbf{x}} = \rho \mathbf{f} + \operatorname{div} \mathbf{T}, \tag{1}$$

$$\rho \left\{ \kappa(\mathbf{U}) \ddot{\mathbf{U}} + \left[\frac{d\kappa}{d\mathbf{U}}(\mathbf{U}) \cdot \dot{\mathbf{U}} \right] \dot{\mathbf{U}} - \frac{1}{2} \frac{d\kappa}{d\mathbf{U}}(\mathbf{U})(\dot{\mathbf{U}} \cdot \dot{\mathbf{U}}) \right\} =$$

$$= \rho \mathbf{B} - \mathbf{Z} + \operatorname{div} \mathbf{\Sigma}, \tag{2}$$

$$\operatorname{skw} \mathbf{T} = 2 \operatorname{skw} \left[\mathbf{U}\mathbf{Z} + (\operatorname{grad} \mathbf{U}) \odot \mathbf{\Sigma} \right], \tag{3}$$

$$\rho \dot{\epsilon} = \mathbf{P} \cdot \dot{\mathbf{F}} + \mathbf{Z} \cdot \dot{\mathbf{U}} + \mathbf{\Sigma} \cdot \operatorname{grad} \dot{\mathbf{U}} + \rho \lambda - \operatorname{div} \mathbf{q}, \tag{4}$$

$$\rho \left(\dot{\psi} + \dot{\theta} \eta \right) \leq \mathbf{P} \cdot \dot{\mathbf{F}} + \mathbf{Z} \cdot \dot{\mathbf{U}} + \mathbf{\Sigma} \cdot \operatorname{grad} \dot{\mathbf{U}} - \theta^{-1} \mathbf{q} \cdot \operatorname{grad} \theta, \tag{5}$$

where $\operatorname{div}(\cdot)$ indicates the divergence of the field $(\cdot)$ and $\operatorname{skw}(\cdot)$ the skew part of a second-order tensor, while the tensor product $\odot$ between third-order tensors is so defined, in components: $(\operatorname{grad} \mathbf{U} \odot \mathbf{\Sigma})_{ij} := \mathbf{U}_{ih,L}\Sigma_{jhL}$.

In the system (1)–(5) we introduced the mass density ρ and its referential value ρ_*; the vector body force $\mathbf{f}$; the Cauchy stress tensor $\mathbf{T}$; the resultant symmetric tensors of external and internal microactions $\rho \mathbf{B}$ and $-\mathbf{Z}$, respectively; the third-order microstress tensor $\mathbf{\Sigma}$, symmetric in the first two places; the densities of internal energy ϵ, of entropy η and of free energy ψ; the rate of heat generation λ due to irradiation or heating supply; the heating flux $\mathbf{q}$.

Finally, on the left hand side of the balance of micro-momentum (2) the Lagrangian derivative of the kinetic co-energy appears, while, on the right hand side, internal microactions $\mathbf{Z}$ include interactive forces between the gross and fine structures as well as internal dissipative contributions due to the stir of the pores' surface, external ones $\mathbf{B}$ are interpreted as controlled pore pressures and the microstress tensor $\mathbf{\Sigma}$ is related to boundary microtractions, even if, in some cases, it could express weakly non-local internal effects.

Remark. The voids theories [13, 2] are recovered by imposing that $\mathbf{U}$ is constrained to be spherical (see, also, §5 of [6]).

3 Constitutive Relations for the Linear Theory

The linear theory of porous solids deals with small changes from the reference placement $\mathcal{B}_*$, that is homogeneous, free of residual macro- and micro-stresses and with zero heat flux rate; the independent kinematic variables are the field of displacement $\mathbf{u}$, the microstrain tensor $\mathbf{V}$ and the temperature change ϑ:

$$\mathbf{u} := \mathbf{x} - \mathbf{x}_*, \quad \mathbf{V} := 2^{-1}(\mathbf{U} - \mathbf{I}) \quad \text{and} \quad \vartheta := \theta - \theta_*, \qquad (6)$$

where $\mathbf{I}$ is the identity tensor.

The constitutive equations for a linear isotropic thermoelastic porous solids, which possess a center of symmetry, relate the stress tensor $\mathbf{T}$, the internal microactions $\mathbf{Z}$, the microstress tensor $\mathbf{\Sigma}$, the heating flux $\mathbf{q}$ and the entropy η to the following measures of "small" thermoelastic deformations: the kinematic variables (6), the infinitesimal strain tensor $\mathbf{E}$ ($:= \operatorname{sym}(\operatorname{grad}\mathbf{u})$, where $\operatorname{sym}(\cdot)$ indicates the symmetric part of a second-order tensor), the gradients of the temperature change $\operatorname{grad}\vartheta$ and of the microstrain $\operatorname{grad}\mathbf{V}$ and its time rate $\dot{\mathbf{V}}$. We observe that this time rate accounts for inelastic surface effects associated with changes in the deformation of pores in the vicinity of void boundaries. Therefore, we derive from [5] (see, also, [7]):

$$\mathbf{T} = \rho_* \left\{ \left[(v_l^2 - 2v_t^2)\operatorname{tr}\mathbf{E} + \kappa_*\lambda_5\operatorname{tr}\mathbf{V} + \gamma_2\vartheta \right]\mathbf{I} + 2v_t^2\mathbf{E} + \kappa_*\lambda_6\mathbf{V} \right\},$$

$$\mathbf{Z} = \rho_*\kappa_* \left\{ \left[\operatorname{tr}\left(\lambda_3\mathbf{V} + \lambda_5\mathbf{E} + \omega\dot{\mathbf{V}} \right) + \gamma_3\vartheta \right]\mathbf{I} + 2\lambda_4\mathbf{V} + \lambda_6\mathbf{E} + 2\sigma\dot{\mathbf{V}} \right\},$$

$$\mathbf{\Sigma} = \rho_*\kappa_*\{\lambda_1\left[\mathbf{I} \otimes \operatorname{div}\mathbf{V} + \operatorname{syml}\left(\operatorname{grad}(\operatorname{tr}\mathbf{V}) \otimes \mathbf{I}\right)\right] + \qquad (7)$$

$$+\lambda_8\mathbf{I} \otimes \operatorname{grad}(\operatorname{tr}\mathbf{V}) + 2\left(v_{tm}^2 - v_{sm}^2\right)\operatorname{syml}\left(\operatorname{div}\mathbf{V} \otimes \mathbf{I}\right) + 2\,v_{sm}^2\operatorname{grad}\mathbf{V}\},$$

$$\mathbf{q} = -\chi\operatorname{grad}\vartheta, \qquad \eta = -\gamma_1\vartheta - \gamma_2\operatorname{tr}\mathbf{E} - \kappa_*\gamma_3\operatorname{tr}\mathbf{V},$$

where $\kappa_* := 2\,\kappa(\mathbf{I})$ is the non-negative microinertia coefficient and the symbols "tr" and "syml" mean, respectively: $\operatorname{tr}\mathbf{D} := \mathbf{D} \cdot \mathbf{I}$, $\forall$ tensor $\mathbf{D}$, and $(\operatorname{syml}\mathbf{\Omega})_{ijl} := \frac{1}{2}\left(\Omega_{ijl} + \Omega_{jil}\right)$, $\forall$ third-order tensor $\mathbf{\Omega}$. The thermoelastic and inelastic constants have to satisfy the following inequalities (see [7]):

$$3v_l^2 > 4v_t^2 > 0, (3v_l^2 - 4v_t^2)(3\,\lambda_3 + 2\,\lambda_4) > \kappa_*(3\,\lambda_5 + \lambda_6)^2, 4\,\lambda_4 v_t^2 > \kappa_*\lambda_6^2,$$

$$v_{tm}^2 > v_{sm}^2 > 0, \; v_{sm}^2\left(6\lambda_1 + 9\lambda_8 + 2v_{tm}^2 + v_{sm}^2\right) > 4(v_{tm}^2 - v_{sm}^2)^2,$$

$$\gamma_1\left[\rho_*\kappa_*^{-1}(3v_l^2 - 4v_t^2)(3\,\lambda_3 + 2\,\lambda_4) - (3\,\lambda_5 + \lambda_6)^2\right] + 6\gamma_2\gamma_3(3\,\lambda_5 + \lambda_6) >$$

$$> 3\gamma_2^2\kappa_*^{-1}(3\,\lambda_3 + 2\,\lambda_4) + 3\gamma_3^2(3v_l^2 - 4v_t^2), \; \gamma_1(3v_l^2 - 4v_t^2) > 3\gamma_2^2, \quad (8)$$

$$\gamma_1 > 0, \; \chi \geq 0, \; \sigma \geq 0, \; 3\omega + 2\sigma \geq 0,$$

where the last three derive from the dissipation inequality gained from (5).

Remark. If the pores were absent, v_l and v_t are recognized to be the propagation speeds of dilatational and distortional waves in the linear isothermal elasticity, respectively, while χ is the usual thermal conductivity.

The field equations governing the displacement field $\mathbf{u}$, the microstrain tensor $\mathbf{V}$ and the temperature change ϑ are obtained by substituting the constitutive relations (7) into the balances $(1)_2$, (2) and (4) as

$$\ddot{\mathbf{u}} = v_t^2\Delta\mathbf{u} + \operatorname{grad}\left[(v_l^2 - v_t^2)\operatorname{div}\mathbf{u} + \kappa_*\lambda_5\operatorname{tr}\mathbf{V} + \gamma_2\vartheta\right] + \kappa_*\lambda_6\operatorname{div}\mathbf{V} + \mathbf{f}, (9)$$

$$\ddot{\mathbf{V}} = v_{sm}^2\,\Delta\mathbf{V} + 2(v_{tm}^2 - v_{sm}^2)\,\mathrm{sym}\,[\mathrm{grad}\,(\mathrm{div}\,\mathbf{V})] + \lambda_1\mathrm{grad}^2(\mathrm{tr}\,\mathbf{V}) + \tag{10}$$

$$+ \left[\lambda_1\mathrm{div}\,(\mathrm{div}\,\mathbf{V}) + \lambda_8\Delta(\mathrm{tr}\mathbf{V}) - \mathrm{tr}(\lambda_3\mathbf{V} + \omega\dot{\mathbf{V}}) - \lambda_5\mathrm{div}\,\mathbf{u} - \gamma_3\vartheta\right]\mathbf{I} -$$

$$-\lambda_6\,\mathrm{sym}\,(\mathrm{grad}\,\mathbf{u}) - 2\lambda_4\mathbf{V} - 2\sigma\dot{\mathbf{V}} + \kappa_*^{-1}\mathbf{B} \qquad \text{and}$$

$$0 = \chi\Delta\vartheta + \gamma_1\dot{\vartheta} + \gamma_2\mathrm{div}\,\dot{\mathbf{u}} + \kappa_*\gamma_3\mathrm{tr}\dot{\mathbf{V}} + \theta_*^{-1}\lambda. \tag{11}$$

By splitting the micromomentum balance (10) into the spherical and deviatoric parts, we have, respectively:

$$\ddot{\nu} = (v_{sm}^2 + \lambda_1 + 3\lambda_8)\Delta\nu + \left[2(v_{tm}^2 - v_{sm}^2) + 3\lambda_1\right]\mathrm{div}\,(\mathrm{div}\,\mathbf{V}) - \tag{12}$$

$$-(3\lambda_3 + 2\lambda_4)\,\nu - (3\lambda_5 + \lambda_6)\mathrm{div}\,\mathbf{u} - (2\sigma + 3\omega)\dot{\nu} - 3\gamma_3\vartheta + \kappa_*^{-1}\iota,$$

$$\ddot{\mathbf{V}}^D = v_{sm}^2\,\Delta\mathbf{V}^D + 2(v_{tm}^2 - v_{sm}^2)\,\mathrm{sym}\,[\mathrm{grad}\,(\mathrm{div}\,\mathbf{V})]^D + \tag{13}$$

$$+\lambda_1(\mathrm{grad}^2\nu)^D - \lambda_6\,\mathrm{sym}\,(\mathrm{grad}\,\mathbf{u})^D - 2\lambda_4\mathbf{V}^D - 2\sigma\dot{\mathbf{V}}^D + \kappa_*^{-1}\mathbf{B}^D,$$

where ν and ι are the traces of $\mathbf{V}$ and $\mathbf{B}$, respectively, while the deviatoric part is defined by: $\mathbf{A}^D := \mathbf{A} - 3^{-1}(\mathrm{tr}\mathbf{A})\mathbf{I}$, $\forall$ 2nd–order tensor $\mathbf{A}$.

4 Small-Amplitude Acoustic Waves

Here we investigate the propagation of small-amplitude acoustic waves in absence of body forces, *i.e.*, $\mathbf{f} = \mathbf{0}$ and $\mathbf{B} = \mathbf{O}$. Therefore, we consider plane wave solutions of the form:

$$\mathbf{u} = \phi(\mathbf{x},\tau)\,\mathbf{h}, \quad \nu = \xi\,\phi(\mathbf{x},\tau), \quad \mathbf{V}^D = \phi(\mathbf{x},\tau)\,\mathbf{W}, \quad \vartheta = \bar{\vartheta}\,\phi(\mathbf{x},\tau), \tag{14}$$

where $\phi(\mathbf{x},\tau) = \mathrm{Re}\,[\exp(ib\tau - \psi\,\mathbf{n}\cdot\mathbf{x})]$ with the wave number $\psi = a + ib/c$; ξ and $\bar{\vartheta}$ are scalar wave amplitudes, $\mathbf{h}$ and $\mathbf{W}$ vector and deviatoric tensor wave amplitudes, respectively; $\hat{\mathbf{h}}$ and $\mathbf{n}$ are unit vectors representing the directions of displacement and of wave propagation, respectively; $b > 0$ is the frequency; $a(b) > 0$ and $c(b)$ are the wave attenuation and the wave speed, respectively; the specific loss is defined by: $l := 4\pi|ac/b|$.

Hence, by inserting (14) in the linear system of balance equations (9), (12), (13) and (11), we obtain the following relations:

$$(b^2 + v_t^2\psi^2)\mathbf{h} + (v_l^2 - v_t^2)\psi^2(\mathbf{h}\cdot\mathbf{n})\mathbf{n} -$$

$$-\kappa_*\psi\left[\lambda_6\mathbf{W}\mathbf{n} + \frac{1}{3}(\lambda_6 + 3\lambda_5)\,\xi\mathbf{n}\right] - \gamma_2\psi\bar{\vartheta}\mathbf{n} = \mathbf{0}, \tag{15}$$

$$\left\{b^2 + \left[\frac{1}{3}\left(2v_{tm}^2 - v_{sm}^2\right) + 2\lambda_1 + 3\lambda_8\right]\psi^2 - 2\lambda_4 - 3\lambda_3 - ib(2\sigma + 3\omega)\right\}\xi +$$

$$+ \left[2(v_{tm}^2 - v_{sm}^2) + 3\lambda_1\right]\psi^2\mathbf{W} + (\lambda_6 + 3\lambda_5)\,\psi(\mathbf{h}\cdot\mathbf{n}) - 3\gamma_3\bar{\vartheta} = 0, \quad (16)$$

$$\left(b^2 + v_{sm}^2\psi^2 - 2\lambda_4 - 2i\sigma b\right)\mathbf{W} + \lambda_6\psi\left[\mathrm{sym}\,(\mathbf{h}\otimes\mathbf{n}) - \frac{1}{3}(\mathbf{h}\cdot\mathbf{n})\mathbf{I}\right] +$$

$$+2(v_{tm}^2 - v_{sm}^2)\psi^2\left[\mathrm{sym}\,(\mathbf{Wn}\otimes\mathbf{n}) - \frac{1}{3}(\mathbf{n}\cdot\mathbf{Wn})\mathbf{I}\right] + \qquad (17)$$

$$+ \left[\frac{2}{3}(v_{tm}^2 - v_{sm}^2) + \lambda_1\right]\psi^2\xi\left[(\mathbf{n}\otimes\mathbf{n}) - \frac{1}{3}\mathbf{I}\right] = \mathbf{O},$$

$$(i\chi\psi^2 - b\gamma_1)\bar{\vartheta} + \gamma_2 b\psi\mathbf{h}\cdot\mathbf{n} - \kappa_*\gamma_3 b\nu = 0. \qquad (18)$$

By means of linear combinations, these algebraic system may be decomposed into five independent systems: two uncoupled equations, two coupled systems of two equations each and one coupled system of four equations.

To study each system, we must introduce two unit vectors, $\mathbf{e}$ and $\mathbf{f}$, in the plane orthogonal to the direction of propagation $\mathbf{n}$ and such that $\mathbf{e}\cdot\mathbf{f} = 0$. Therefore, we have the following particular cases.

4.1 Shear Optical Waves

From equation (17), we obtain two different shear optical waves:

$$\left(b^2 + v_{sm}^2\psi^2 - 2\lambda_4 - 2i\sigma b\right)W_{ef} = 0 \quad \text{and} \qquad (19)$$

$$\left(b^2 + v_{sm}^2\psi^2 - 2\lambda_4 - 2i\sigma b\right)(W_{ee} - W_{ff}) = 0, \qquad (20)$$

where the subscripts indicates components. They propagates with attenuation $a = \frac{\sigma c}{v_{sm}^2}$ and speed given by $c^2 = \frac{v_{sm}^2}{2\sigma^2}\left[2\lambda_4 - b^2 + \sqrt{(2\lambda_4 - b^2)^2 + 4\,\sigma^2 b^2}\right]$ without modifying the thermo–elastic features of the matrix material of the porous medium; then the specific loss is $4\pi\left|\frac{2\lambda_4 - b^2}{2b\sigma} + \sqrt{1 + \left(\frac{2\lambda_4 - b^2}{2b\sigma}\right)^2}\right|$.

For low frequencies, the speed and the attenuation approach $v_{sm}\sqrt{2\lambda_4}/\sigma$ and $\sqrt{2\lambda_4}/v_{sm}$, respectively, while l is big; for high frequencies all quantities grow with b.

Moreover, it is also possible a static solution with attenuation $a = \frac{\sqrt{2\lambda_4}}{v_{sm}}$.

4.2 Transverse Waves

From equations (15) and (17), we obtain two different systems of transverse waves, for $j = e, f$:

$$(b^2 + v_t^2\psi^2)h_j - \kappa_*\lambda_6\psi W_{nj} = 0, \tag{21}$$

$$\lambda_6\psi h_j + 2\left(b^2 + v_{tm}^2\psi^2 - 2\lambda_4 - 2i\sigma b\right)W_{nj} = 0. \tag{22}$$

A necessary and sufficient condition for these two homogeneous systems to have a nontrivial solution for the amplitudes h_j and W_{nj} is that the dispersion relation is satisfied by ψ:

$$2\left(v_t^2\psi^2 + b^2\right)\left(v_{tm}^2\psi^2 + b^2 - 2\lambda_4 - 2i\sigma b\right) + \kappa_*\lambda_6^2\psi^2 = 0. \tag{23}$$

This equation is similar to the dispersion relation for plane thermoelastic waves studied in [15] and our analysis will follow the one there developed. One transverse solution of (23) is associated predominantly with the elastic properties of the material (v_t) and denoted by ψ_t; the other ψ_{tm} with the properties governing changes in porosity $(v_{tm}, \lambda_4, \lambda_6$ and $\sigma)$.

The exact solutions of the dispersion relation (23) are complex, and are summarized in Table 1 of [16] (*modulo* some innocuous identification in notations), but, in our context, they can be interpreted physically without a great deal of difficulty. The coupling of motion equations (15) and (17) of linear macro– and micro–momentum makes the wave of dispersive kind, while the presence of big pores adds a dissipative mechanism associated with voids that causes both waves to attenuate. The resonance is present only if the dissipation coefficient σ is null.

At high frequencies, the predominantly elastic transverse wave propagates with the classical speed v_t and, as the frequency approaches infinity, the attenuation coefficient a and the specific loss l are very small and approach zero. Instead the predominantly micro-transverse wave propagates with constant speed v_{tm} and attenuation σ/v_{tm}, but with a small specific loss which approach zero when the frequency approaches infinity.

At low frequencies, the elastic wave propagates with speed $v_t\sqrt{1-\nu}$, where $0 \leq \nu \left(:= \frac{\kappa_*\lambda_6^2}{4\lambda_4 v_t^2}\right) < 1$ for the inequality $(8)_4$, while, as for large frequencies, the attenuation coefficient and the specific loss remain very small and approach zero with the frequency itself. The predominantly micro-transverse wave propagates with constant speed $\frac{v_{tm}}{\sigma}\sqrt{2\lambda_4(1-\nu)}$ and constant attenuation $\frac{\sqrt{2\lambda_4(1-\nu)}}{v_{tm}}$; however, the specific loss $l = \frac{2\lambda_4(1-\nu)}{\sigma b}$ is large and in inverse proportion to the frequency b.

The amplitude ratio R of the micro–wave to the macro–wave is obtained from (21): $R = W_{nj}/h_j = (b^2 + v_t^2\psi^2)/\kappa_*\lambda_6\psi$. For the solution ψ_t at large frequencies, the ratio R_t is a constant; at small frequencies it approaches zero with the frequency itself. Instead, the micro–mode ψ_{tm} at large frequencies gives a ratio R_{tm} very big, while at low frequencies it is constant.

Finally, we have a static solution with attenuation $a = \frac{\sqrt{2\lambda_4(1-\nu)}}{v_{tm}}$ in which the amplitudes are related by $h_j = \frac{\kappa_*\lambda_6}{v_t^2 a}W_{nj}$, for $j = e, f$.

4.3 Longitudinal Waves

At the end we obtain longitudinal waves from the remaining equations of the system (15)–(18):

$$(b^2 + v_l^2\psi^2)h_n - \kappa_*\psi\left[\lambda_6 W_{nn} + \frac{1}{3}(\lambda_6 + 3\lambda_5)\,\xi\right] - \gamma_2\psi\bar{\vartheta} = 0, \tag{24}$$

$$\left[b^2 + \left(\frac{2}{3}v_{tm}^2 - \frac{1}{3}v_{sm}^2 + 2\lambda_1 + 3\lambda_8\right)\psi^2 - 2\lambda_4 - 3\lambda_3 - ib(2\sigma + 3\omega)\right]\xi +$$

$$+ (\lambda_6 + 3\lambda_5)\,\psi h_n + \left[\frac{2}{3}(v_{tm}^2 - v_{sm}^2) + \lambda_1\right]\psi^2 W_{nn} - 3\gamma_3\bar{\vartheta} = 0, \tag{25}$$

$$\frac{2}{3}\lambda_6\psi h_n + \left[b^2 + \frac{1}{3}\left(4v_{tm}^2 - v_{sm}^2\right)\psi^2 - 2\lambda_4 - 2i\sigma b\right]W_{nn} + \tag{26}$$

$$+ \frac{2}{3}\left[\frac{2}{3}(v_{tm}^2 - v_{sm}^2) + \lambda_1\right]\psi^2\xi = 0,$$

$$\gamma_2 b\psi h_n - \kappa_*\gamma_3 b\xi + (i\chi\psi^2 - b\gamma_1)\bar{\vartheta} = 0. \tag{27}$$

To have a nontrivial solution for the amplitudes h_n, ξ, W_n and $\bar{\vartheta}$ we have to set the determinant of their coefficients equal to zero and we obtain a 4th–order equation in ψ^2, which is been resolved with the Ferrari–Cardano derivation of the quartic formula, after the application of the Tchirnhaus transformation. The exact analytical solutions of the dispersion relation for longitudinal waves are too complicated and long to be reported here explicitly, and however the solution can best be undertaken by numerical techniques: they will be presented, together with numerical resolutions, in a forthcoming work.

The four solutions ψ_e^2, ψ_d^2, ψ_v^2 and ψ_{th}^2 for the wave numbers are dominated by displacement, deviatoric and spherical part of microstrain and thermal fields, respectively, and show the existence of four coupled compressional waves: the first one ψ_e^2 is predominantly an elastic wave of dilatation, the second one ψ_d^2 is associated with an equivoluminal microelastic wave, the third one ψ_v^2 is predominantly a volume fraction wave of pure dilatation, the last one corresponding to ψ_{th}^2 is similar in character to a thermal wave. Let us observe that, when we solve the quartic formula by using a computer program, the micro-waves are slower than elastic and thermal waves, the elastic one being the fastest: this is in accordance with experimental evidence.

We observe that, if we neglect thermal effects, our micro-wave solutions above can be recognized in some developments of §8 of [12] for elastic plates, if micro-rotations are disregarded. In particular, the velocity of the elastic wave is less than that which would be calculated for classical elasticity v_l and the phenomenon is due to the compliance of pores. Instead, if we neglect non-spherical contributions of the microstrain to constitutive equations (7), we recover the solutions of voids theory [17].

Also now the coupling of all the equations (24) and (27) makes the longitudinal wave dispersive in character and, because of the thermal coupling and of the presence of voids, suffering attenuation (the thermal mode with a large coefficient). Moreover, the resonance is present only if the two dissipation coefficient σ and ω are null.

At low frequencies, there is no damping effect in either of the four modes. The velocity v_l of the ψ_e–wave is increased by a small amount due to the thermomechanical coupling, but decreased significantly because of porosity effects; the attenuation is a quite small constant. The other three modes almost do not exist and their attenuation coefficients remain very small and approach zero with the frequency itself.

At high frequencies, the ψ_v–wave, which travels with speed $v_v^2 = \frac{5}{3}v_{sm}^2 - \frac{2}{3}v_{tm}^2 + 3\lambda_8 > 0$ for $(8)_{4,5}$, and the ψ_d one, of velocity $v_d^2 = 2v_{tm}^2 - v_{sm}^2 + \lambda_1 > 0$ by inspection, are not accompanied by thermal or elastic modes, which are instead coupled, and *viceversa*; attenuation coefficients for micro–modes remains small but constant. The speed of propagation and the attenuation coefficient of the thermal mode ψ_{th} sharply increase with the frequency itself, being diffusive in nature, instead elastic wave propagates with the classical speed v_l and attenuation coefficient which approaches zero slowly with the frequency itself.

We notice that the big frequency limits of the two micro–elastic waves corresponds to the velocities of acceleration waves in the same material, undeformed and at rest [8].

It is observed numerically that the specific loss l is significantly large when the wave speed has quite small value in various regions of frequency. The specific loss due to energy dissipation is comparatively high in case of ψ_e and ψ_{th}–waves and moderate for predominantly micro–elastic waves of dilatation.

Acknowledgements. Work supported by the Department of Mechanics and Materials, Faculty of Engineering, "Mediterranean" University of Reggio Calabria, Italy.

References

1. Capriz, G., Giovine, P.: Math. Mod. Meth. Appl. Sciences 7, 211–216 (1997)
2. Capriz, G., Podio-Guidugli, P.: Arch. Rational Mech. Anal. 75, 269–279 (1981)
3. Cieszko, M.: Extended Description of Pore-Space Structure and Fluid Flow in Porous Materials. Application of Minkowski Space. In: Proc. of the XIVth International STAMM 2004, pp. 93–102. Shaker Verlag, Aachen (2004)
4. Cowin, S.C.: Bone 22(suppl.), 119S–125S (1998)
5. Eringen, A., Suhubi, E.: Int. J. Eng. Sci. 2, 189–203, 389–404 (1964)
6. Giovine, P.: Porous Solids as Materials with Ellipsoidal Structure. In: Batra, R.C., Beatty, M.F. (eds.) Contemporary Research in the Mechanics and Mathematics of Materials, pp. 335–342. CIMNE, Barcelona (1996)
7. Giovine, P.: Trans. Porous Media 34, 305–318 (1999)
8. Giovine, P.: On Acceleration Waves in Continua with Large Pores. In: Proc. of the XIVth International STAMM 2004, pp. 113–124. Shaker Verlag, Aachen (2005)

9. Grioli, G.: Continuum Mech. Thermodyn. 15, 441–450 (2003)
10. Hasselman, D.P.H., Singh, J.P.: Criteria for Thermal Stress Failure of Brittle Structural Ceramics. In: Thermal Stresses I, Northolland, Amsterdam (1982)
11. Lakes, R.S.: Int. J. Solids Structures 22, 55–63 (1986)
12. Mindlin, R.D.: Arch. Rational Mech. Anal. 16, 51–78 (1964)
13. Nunziato, J.W., Cowin, S.C.: Arch. Rational Mech. Anal. 72, 175–201 (1979)
14. Ponte-Castañeda, P., Zaidman, M.: J. Mech. Phys. Solids 42, 1459–1497 (1994)
15. Puri, P.: Int. J. Eng. Sci. 10, 467–477 (1972)
16. Puri, P., Cowin, S.C.: J. of Elasticity 15, 167–183 (1985)
17. Singh, J., Tomar, S.K.: Mechanics of Materials 39, 932–940 (2007)

Viscoelastic Modeling of Brain Tissue: A Fractional Calculus-Based Approach

Vincent Libertiaux and Frédéric Pascon

Abstract. In recent years, the mechanical study of the brain has become a major topic in the field of biomechanics. A global biomechanical model of the brain could find applications in neurosurgery and haptic device design. It would also be useful for car makers, who could then evaluate the possible trauma due to impact. Such a model requires the design of suitable constitutive laws for the different tissues that compose the brain (i.e. for white and for gray matters, among others).

Numerous constitutive equations have already been proposed, based on linear elasticity, hyperelasticity, viscoelasticity and poroelasticity. Regarding the strong strain-rate dependence of the brain's mechanical behaviour, we decided to describe the brain as a viscoelastic medium. The design of the constitutive law was based on the Caputo fractional derivation operator. By definition, it is a suitable tool for modeling hereditary materials. Indeed, unlike integer order derivatives, fractional (or real order) operators are non-local, which means they take the whole history of the function into account when computing the derivative at current time t.

The model was calibrated using experimental data on simple compression tests performed by Miller and Chinzei. A simulated annealing algorithm was used to ensure that the global optimum was found. The fractional calculus-based model shows a significant improvement compared to existing models. This model fits the experimental curves almost perfectly for natural strains up to -0.3 and for strain-rates from $0.64 s^{-1}$ to $0.64\,10^{-2}\,s^{-1}$.

Vincent Libertiaux and Frédéric Pascon
University of Liège, Department of Architecture,
Geology, Environment and Constructions - Chemin des Chevreuils,
1 4000 Liège - Belgium
e-mail: `vincent.libertiaux@ulg.ac.be,f.pascon@ulg.ac.be`

J.-F. Ganghoffer, F. Pastrone (Eds.): Mech. of Microstru. Solids, LNACM 46, pp. 81–90.
springerlink.com

Notations

$D_t^{(\alpha)}$	Real-order derivative with respect to t
$\mathbf{S}$	Second Piola-Kirchhoff stress tensor
$\mathbf{F}$	Deformation gradient tensor
$\mathbf{C} = \mathbf{F}^T\mathbf{F}$	Left Cauchy Green tensor
$\mathbf{g}$	Boldface latin letter: tensor quantity
L_1	Set of summable functions

1 Introduction

Although the field of brain biomechanics has been investigated for more than 30 years, theris still no consensus within the scientific community yet as to the optimal model for brain tissues. All agree that brain tissues do not behave like elastic solids, though such constitutive laws are used in some works (5),(10),(19), but there are two conflicting ways of thinking, both presenting advantages and drawbacks. These are the poroelastic and the viscoelastic approaches.

The poroelastic theory comes from soil mechanics. Indeed, soils are made of porous rocks (partially) filled with water. Since brain is composed of hydrated tissues, authors like Miga (11),(12) prefer this approach. According to him, there is indubitably viscoelasticity in brain tissues but a pure viscoelastic description would be limited given the inherent coupling between deformation and hydrodynamic behavior. Using a 20,000 nodes geometrically realistic mesh of the brain, Miga & al achieved a 75 to 85 percent predictive capability on the displacements of 20 beads inserted in a pig's brain during a *in vivo* compression test (11).

The literature is more abundant as far as viscoelastic models are concerned. Their strain rate dependence make them more likely to model the wide range of behaviors encountered as functions of the loading velocity. Unlike poroelastic models which focus on neurosurgical applications only, viscoelastic models have been used in several domains. Sarron & al (18) have designed a 2D multi-domain model to study the G loss of consciousness (GLOC) of fighter pilots. Brands & al (3) focused on the design of a 3D non linear viscoelastic constitutive model for brain tissue during impact. They used a differential decoupled 3D constitutive law. Darvish and Crandall (6) agreed with the nonlinear viscoelastic behavior of the brain when subjected to deformation impulses with duration of only a few milliseconds. According to them, the nonlinear effects in the brain constitutive relation are of particular importance for two reasons:

1. a non linear model is more accurate for finite deformations,
2. the injury pattern predicted by a non linear model can potentially differ from that of a linear one. According to linear theory of materials with memory, the waves produced by a step-wise input are diffusive in character, broadening with time, whereas the non linear theory permits

the development of discontinuities in the form of shock or acceleration waves.

Shear tests showed a strong non linearity behavior, even for strain as small as 1%.

Contrarily to the previous authors, Velardi (20) claims that viscous effects can be disregarded in impact loading as they have a limited influence on the short term response of the brain tissue. The model used by Velardi is hyperelastic and takes the anisotropy of the white matter into account.

Miller designed hyperviscoelastic models (15),(13) valid over a wide range of strain rates in traction and compression. His constitutive equation was implemented in a FE model of human brain. The aim was to study the displacements due to the brain shift at the opening of the skull (21).

2 Fractional Calculus

The concept of fractional (or real-order) derivative is not new, as it was first dicussed by l'Hospital and Leibniz in 1695. Nevertheless, it may sound weird the first time it is heard. Indeed, it is nor usual do deal with an expression such as $\left(\dfrac{d}{dt}\right)^{1.4} f(t)$ neither it is to determine its physical meaning, if there is one.

Amongst the suitable definitions that fit the concept of real-order derivative, the Riemann-Liouville and Grunwald-Letnikov's formulations are the most famous:

Riemann-Liouville

$$D_t^{(\alpha)} f(t) = \frac{1}{\Gamma(-\alpha)} \int_0^t \frac{f(x)}{(t-x)^{(1+\alpha)}} \, dx \tag{1}$$

Grunwald-Letnikov

$$D_t^{(\alpha)} f(t) = \lim_{h \to 0} \left[h^{-\alpha} \sum_{m=0}^{\frac{t}{h}} (-1)^m \frac{\Gamma(\alpha+1)}{\Gamma(\alpha-m+1)\Gamma(m+1)} f(t-mh) \right] \tag{2}$$

where $\Gamma(x)$ refers to the Eulerian gamma function:

$$\Gamma(x) = \int_0^\infty e^{-t} t^{x-1} \, dt \tag{3}$$

According to (16), both definitions are equivalent for the fractional index $p, 0 < p < n$, if the function $f(t)$ is $(n-1)$-times continuously differentiable in the interval $[0, T]$ and if $f^{(n)}(t)$ is integrable in $[0, T]$. This property proves to be useful as definition (2) is more convenient for numerical evaluation and (1) for analytical calculus.

It appears directly that the main difference between classical integer order derivative operators and fractional ones is that the latter is non local : all the past values of the function f are required to compute $D^{(\alpha)}f$ at time t.

Fractional derivation operators seem thus a good tool to model hereditary materials. Furthermore, the convolution kernel $(t - .)^{-(1+\alpha)}$ in expression 1 exhibits the behaviour of a fading memory function as it is shown in figure 1. For decades, the Riemann operator played a major role in the development of

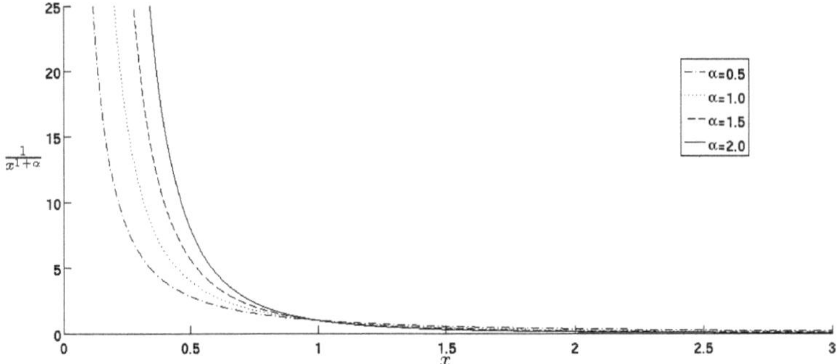

Fig. 1 Convolution kernel of the Riemann fractional derivative for several α

the theory of fractional derivatives and for its application in the pure mathematical field (solution of differential equations, definition of a new function classes,...). However, since the physical processes have been modeled with the help of fractional derivatives, the theory needed to be revised or at least extended. Indeed, let us examine the real order derivative of a constant $c \in \mathbb{R}$

$$D_t^{(\alpha)}c = \frac{1}{\Gamma(\alpha)} \int_0^t \frac{c}{(t-x)^{1+\alpha}}\, dx \tag{4}$$

$$= \frac{c}{\Gamma(-\alpha)} \frac{t^{-\alpha}}{\alpha} \tag{5}$$

The fractional derivative of a constant is not 0 but rather a real-order power of the independent variable. This result implies that the use of Riemann derivative in the frame of fractional differential equations will lead to impose fractional initial conditions, which do not have an immediate physical interpretation, unlike integer-order ones.

The alternative is then to use another definition of the fractional derivative which returns 0 when deriving a constant. The Caputo derivative has been designed specifically to present this property. Its definition is given by expression 6:

$$^C D_t^{(\alpha)} f(t) = \frac{1}{\Gamma(n-\alpha)} \int_0^t \frac{f^{(n)}(x)}{(t-x)^{\alpha+1-n}}\, dx, \quad n = \lceil \alpha \rceil \tag{6}$$

with $\lceil . \rceil$ meaning *the smallest integer greater than* . One can easily verify that

$$\lim_{\alpha \to n} {}^{C}D_t^{(\alpha)} f(t) = f^{(n)}(t) \tag{7}$$

2.1 Fractional Differential Equations

Unfortunately, the Grunwald's derivative, which is very suitable for the numerical resolution of fractional differential equations (FDE) presents the same drawback (initial conditions) as the Riemann's definition. From the foregoing, we will focus on the Caputo derivative[1] in the framework of FDE.

Let us consider a linear FDE:

$$D_t^{(\alpha_n)} + \sum_{j=1}^{n-1} p_j(t) D_t^{(\alpha_{n-j})} y(t) + p_n(t) y(t) = f(t) \tag{8}$$

with initial conditions $y^{(k)}(0) = y_0^k$, $k = 0, 1, 2, \cdots, n-1$

We must first make sure that a solution exists and is unique. This is the case if the conditions of the following theorem are fulfilled:

> If $f(t) \in L_1]0, T[$, and $p_j(t)$, $(j = 1, 2, ..., n)$ are continuous functions in the closed interval [0,T], then the initial-value problem (8) has a unique solution $y(t) \in L_1]0, T[$

An existence and uniqueness theorem also exists for non linear equations. There will be no further development as we will only deal with linear FDE's.

Fractional constitutive models have been studied on a theoretical point of view by Adolfsson, Enelund and Olsson (2), (1). Freed and Diethelm (8) applied fractional calculus to the modeling of calcaneal fat pad.

3 Fractional Hyperviscoelastic Law

Hyperviscoelastic models have already been used in the framework of brain tissue modeling by Miller (13), (15). The general form of such laws is written

$$\Phi = \int_0^t f_{rel}(t - \tau) \frac{\partial W_h}{\partial \tau} \, d\tau \tag{9}$$

---

[1] The superscript ^{C} will then be omitted.

In the case of (13), expression 9 takes the particular form

$$\Phi \;=\; \int_0^t \left\{ \sum_{i+j=1}^{N} \left[C_{ij0} \left(1 - \sum_{k=1}^{n} g_k (1 - e^{(\tau-t)/t_k}) \right) \right] \frac{\partial}{\partial \tau} \left[(I_1 - 3)^i (I_2 - 3)^j \right] \right\} d\tau \tag{10}$$

where C_{ij0} are hyperelastic coefficients, τ_k relaxation times and g_k relaxation modulus. N is the order of polynomial in strain invariants. n, the number of relaxation terms needed to fit the experimental curves. Miller chose $N = n = 2$ and assumed $C_{ij0} = C_{ji0}$ and $C_{110} = 0$.

The second Piola-Kirchhoff stress tensor, in the case of an incompressible material, is given by:

$$\mathbf{S} = \frac{\partial \Phi}{\partial \mathbf{C}} - p\mathbf{C}^{-1} \tag{11}$$

$$= \mathbf{S}_c - p\mathbf{C}^{-1} \tag{12}$$

where p stands for an unknown hydrostatic pressure to be determined by the boundary conditions and the equilibrium equations. The firs term $\mathbf{S}_c$ is thus

$$\mathbf{S}_c \;=\; \int_0^t \left\{ \sum_{i+j=1}^{2} \left[C_{ij0} \left(1 - \sum_{k=1}^{2} g_k (1 - e^{(\tau-t)/t_k}) \right) \right] \frac{\partial}{\partial \tau} \frac{\partial}{\partial \mathbf{C}} \left[(I_1 - 3)^i (I_2 - 3)^j \right] \right\} d\tau \tag{13}$$

A differential equation equivalent to (13) can be obtained and is written

$$\ddot{\mathbf{S}}_C + \beta \dot{\mathbf{S}}_C + \frac{1}{\tau_1 \tau_2} \mathbf{S}_C = \mathbf{G}(t) \tag{14}$$

provided that

$$\beta = \frac{1}{\tau_1} + \frac{1}{\tau_2} \tag{15}$$

$$\mathbf{G}(t) = \dot{\mathbf{p}}(t) + \frac{1 - g_1 - g_2}{\tau_1 \tau_2} \mathbf{F_p}(t) \tag{16}$$

$$\dot{\mathbf{p}}(t) = \frac{\partial \mathbf{p}(t)}{\partial t} \tag{17}$$

$$\mathbf{F_p}(t) = \int_0^t \mathbf{p}(\tau) \, d\tau \tag{18}$$

$$\mathbf{p}(t) = \sum_{i+j=1}^{2} C_{ij0} \frac{d}{dt} \frac{\partial}{\partial \mathbf{C}} \left((I_1 - 3)^i (I_2 - 3)^j \right) \tag{19}$$

The solution of the homogeneous problem relative to (14) describes the relaxation of the material while the right hand side term is related to the instantaneous applied strain.

According to (16) and (4) the relaxative-oscillative behavior characteristic of second order differential equations can be reproduce using a single fractional-order derivative. The idea is thus to replace equation (14) with an suitable FDE:

$$\mathbf{S}^{\alpha} + b\mathbf{S} = \mathbf{H}(t) \tag{20}$$

The right hand side $\mathbf{H}(t)$ requires a further explanation. Indeed, the expression 16 is not valid anymore as it is dimensionally incorrect in the framework of the FDE. Thus, $\mathbf{H}(t)$ must be written under the form:

$$\mathbf{H}(t) = \eta\dot{\mathbf{p}}(t) + \zeta\mathbf{F_p}(t) \tag{21}$$

as τ_k and g_k ($k = 1, 2$) have no longer any physical signification. The dimensions of ζ and η are respectively $T^{-\alpha}$ and $T^{2-\alpha}$. The fractional constitutive law for brain tissue then writes:

$$\mathbf{S} = \mathbf{S}^C - p\mathbf{C}^{-1}$$
$$D_t^{(\alpha)}\mathbf{S}^C + b\mathbf{S}^C = \mathbf{H}(t) \tag{22}$$

Due to the complexity of the function $\mathbf{H}(t)$, the analytical resolution of equation 20 is not possible. A numerical algorithm must then be used. The implementation that we propose to use is due to Diethelm,Ford and Freed (7) who carried over the essential ideas of the Adams-Moulton method to the fractional order problem.

4   Results

The determination of the six parameters $(\alpha, b, C_{100}, C_{200}, \eta, \zeta)$ was achieved through a stochastic optimization (the simulated annealing algorithm (17)). The experimental data used were the results of Miller's unconfined compression tests at fast and medium loading velocities. The following values were obtained:

α	1.04
b	$8.581\ 10^{-3}$
C_{100}	3463.7
C_{200}	35.908
η	1.0
ζ	$5.042\ 10^{-2}$

Figures 2 and 3 show the curves obtained from the model and the experimental data. The model fits almost perfectly the data for the medium loading velocity, while the prediction for the high loading velocity is very good up to a natural deformation equals to -0.3.

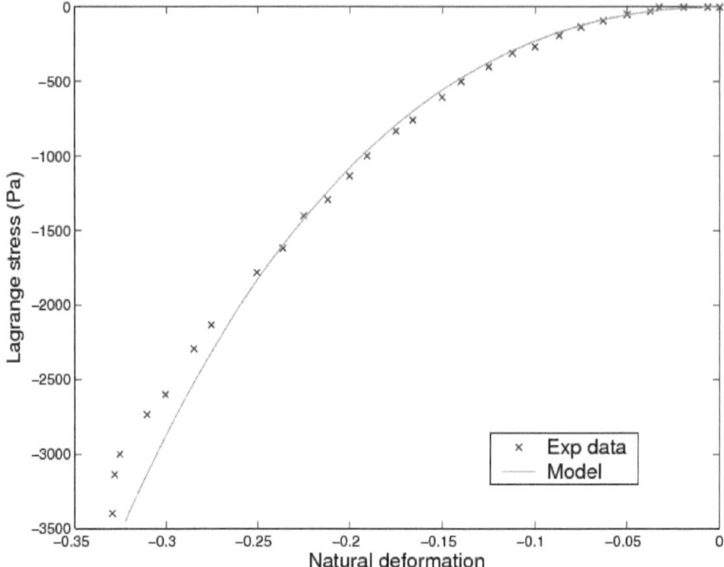

Fig. 2 Lagrange stress vs natural strain for the unconfined compression experiment at strain rate $\dot{\varepsilon} = 0.64s^{-1}$

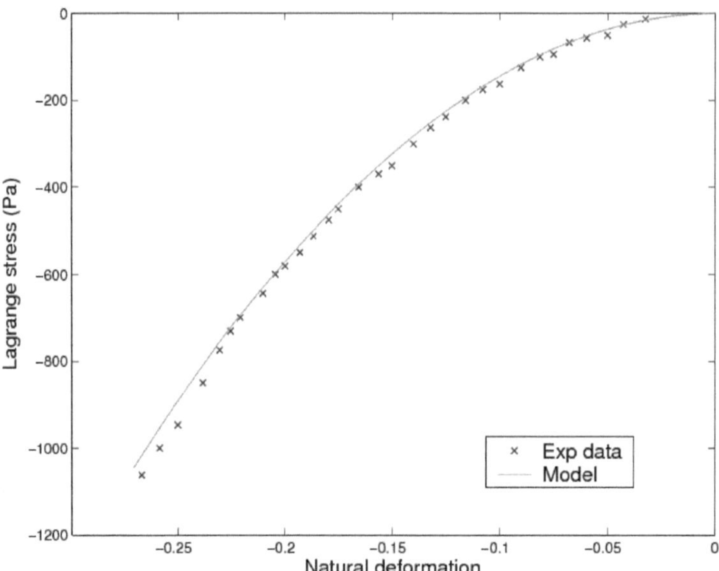

Fig. 3 Lagrange stress vs natural strain for the unconfined compression experiment at strain rate $\dot{\varepsilon} = 0.0064s^{-1}$

References

[1] Adolfsson, K.: Models and numerical proecures for fractional order viscoelasticity. PhD thesis, Chalmers university of Technology (2003)

[2] Adolfsson, K., Enelund, M.: Fractional derivative viscoelasticity at large deformations. Nonlinear dynamics 33, 301–321 (2003)

[3] Brands, D., Peters, G., Bovendeed, P.: Design and numerical implementation of a 3d non-linear viscoelastic constitutive model for brain tissue during impact. Journal of Biomechanics 37, 127–134 (2004)

[4] Carpinteri, A., Mainardi, F.: Fractals and fractional calculus in continuum mechanics. Springer, Heidelberg (2006)

[5] Clatz, O.: Analysis and prediction of the brain deformation during a neurosurgical procedure. Master's thesis, ENS Cachan (2002)

[6] Darvish, K., Crandall, J.: Nonlinear viscoelastic effects in oscillatory shear deformation of brain tissue. Medical Engineering & Physics 23, 633–645 (2001)

[7] Diethelm, K., Ford, N.J., Freed, A.D.: A predictor-corrector approach for the numerical solution of fractional differential equations. Non linear dynamics 29, 3–22 (2002)

[8] Diethelm, K., Freed, A.D.: Fractional calculus in biomechanics: a 3d viscoelastic model using regularized fractional-derivative kernels with application to the human calcaneal fat pad. Biomechanics and Modeling in Mechanobiology 5, 203–215 (2006)

[9] Ford, N.J., Simpson, A.C.: The numerical solution of fractional di erential equations: speed versus accuracy. Technical report, Manchester Center for Computational Mathematics (2001)

[10] Hault, A., Drazetic, P., Razafimahery, F.: Etudes des phenomenes d'interaction fluide/structure lors d'un choc a l'interieur de la boite cranienne. In: 17eme Congres Francais de Mecanique (2005)

[11] Miga, M., Paulsen, K., Hoopes, P., Kennedy, F., Hartov, A., Roberts, D.: In vivo modelling of interstitial pressure in the brain under surgical load using finite elements. Journal of Biomechanical Engineering 122, 354–363 (2000)

[12] Miga, M., Paulsen, K., Hoopes, P., Kennedy, F., Hartov, A., Roberts, D.: In vivo quantification of a homogeneous brain deformation model for updating preoperative images during surgery. IEEE Transactions on Biomedical Engineering 47, 266–273 (2000)

[13] Miller, K.: Constitutive model of bain tissue suitable for finite element analysis of surgical procedures. Journal of Biomechanics 32, 531–537 (1999)

[14] Miller, K., Chinzei, K.: Constitutive modelling of brain tissue: experiment and theory. Journal of Biomechanics 30, 1115–1121 (1997)

[15] Miller, K., Chinzei, K.: Mechanical properties of brain tissue in tension. Journal of Biomechanics 35, 483–490 (2002)

[16] Podlubny, I.: Fractional differential equations. London Academic Press, London (1999)
[17] Samuelides, M.: Calcul intensif: optimisation stochastique. Supaero (1995)
[18] Sarron, J.C., Blondeau, C., Guillaume, A., Osmont, D.: Identification of linear viscoelastic constitutive models. Journal of Biomechanics 33, 685–693 (2000)
[19] Skrinjar, O., Nabavi, A., Duncan, J.: Model-driven brain shift compensation. Medical Image Analysis 6, 361–373 (2002)
[20] Velardi, F., Fraternali, F., Angelillo, M.: Anisotropic constitutive equations and experimental tensile behavior of brain tissue. Biomech. Model Mechanbiol. 5(53), 61 (2006)
[21] Wittek, A., Miller, K., Kikinis, R., Warfield, S.: Patient-specific model of brain deformation: application to medical image registration. Journal of Biomechanics 40, 919–929 (2007)

Modelling of the Mechanical Behaviour of Regular Metallic Hollow-Sphere Packings under Compressive Loads

V. Marcadon, E. Roques, and F. Feyel

Abstract. This paper is devoted to the study of the mechanical behaviour under compression of hollow-sphere packings and, especially, to the characterization of the mechanisms that govern their compaction. The aim of this work is to understand how the architecture of such materials governs their overall mechanical behaviour through a combined approach involving both experiments and modelling. In this work, emphasis is put on the influence of local plasticity and buckling of spheres on the effective behaviour of packings. A finite element modelling is developed taking into account those mechanisms in order to characterize the contributions of the stiffness of the meniscuses and the hollow spheres respectively on the effective stiffness of packings.

1 Introduction

Cellular materials seem to be particularly interesting for the development of lightweight aeronautical frames since better specific properties than bulk and multi-functional materials are expected [3]. For example, a cellular material with an optimized architecture, with respect to cell size, shape and arrangement, can simultaneously have good acoustic absorbance properties, structural resistance and impact absorption properties. Honeycombs or metal foams are already used for such applications. The work presented here concerns the study of the potential of an alternative solution: metal hollow-sphere packings (HSPs).

Only a few studies exist on the mechanical behaviour of such materials. Indeed, most of the work done on sphere packings deal with physical and statistical

V. Marcadon, E. Roques, and F. Feyel
Département Matériaux et Structures Métalliques, Office National d'Etudes et de
Recherches Aérospatiales, 29 av. de la Division Leclerc, BP 72,
F-92322 Châtillon Cedex, France

V. Marcadon
Corresponding author. Tel.: +33 146734524, Fax: +33 146734164,
e-mail: Vincent.Marcadon@onera.fr

J.-F. Ganghoffer, F. Pastrone (Eds.): Mech. of Microstru. Solids, LNACM 46, pp. 91–99.
springerlink.com

aspects of these materials in order to characterize the different packing types in terms of compactness, coordination number, *etc.* (Aste *et al.* [1], Müeller *et al.* [8], Scott *et al.* [14,15]). Nevertheless, one can cite the work of Sanders and Gibson [12,13] on the mechanical behaviour of regular HSPs who have shown that their mechanical properties depend strongly on the solicitation direction; they are stiffer along the denser direction of the packing. However, it can be objected that those works only treat the case of small deformations, which seems too restrictive in view of their potential applications. Contrarily to HSPs, metal foams have been widely studied. Since their architectures present some similarities with the one of HSPs, conclusions obtained on these materials may provide relevant information for the study of HSPs. All the studies agree that local plasticity and the collapse of constitutive cells have a significant influence on the mechanical response of foams, see Brothers and Dunand [2] on zirconium foams, Paul and Rammamurty [9] or Ruan *et al.* [11] on aluminium foams. They also agree with the fact that several elastic and plastic mechanical properties of foams vary with their density according to a power law. However, the influence of the solicitation rate on the mechanical response of metal foams is a more debated issue. According to Paul and Rammamurty [9] the foams stiffness increases with the solicitation rate, whereas for Ruan *et al.* [11] and Rakow and Waas [10] no effect of the solicitation rate exists. These studies have been performed on aluminium foams. Concerning the mechanical modelling of metal foams, Gibson and Ashby [4,5] proposed an analytical model to predict the mechanical properties of a foam through the knowledge of the mechanical properties of its constitutive material and densities. This model, developed for both open- and close-cell foams, assumes that the edges of the foam are loaded under flexion. A similar approach has been proposed by Hodge and Dunand [6] assuming that the edges of the foam work in compression.

This paper is devoted to the mechanical characterization under compression of HSPs and to the development of a finite element (FE)-modelling of their mechanical behaviour. The first part of this work presents an experimental characterization of the mechanical behaviour of HSPs under compression and, particularly, of the mechanisms that govern this behaviour. In a second part, this information is introduced in a FE-modelling in order to evaluate the relevance of such an approach to accurately appreciate the compaction behaviour of such materials. A study of the influence of the respective stiffness of the constitutive components of HSPs (*i.e.*, the hollow spheres and the meniscuses respectively) on their effective stiffness is also presented. The third part is dedicated to the comparison between experimental results and modelling predictions.

2 Experimental Characterization of the Compression Behaviour of HSPs

2.1 Samples Description and Test Conditions

Samples for compression tests are cylinders of nickel hollow spheres linked to each other by a nickel-boron braze (fig. 1a). The packing is non-regular and radius and length of the cylinders equal 30 mm and 27 mm respectively. Samples were cut by electro-erosion out of a larger plate of 480×310×27 mm^3, provided by Ateca. The outer radius and thickness of the constitutive hollow spheres approximately equal 3±0.5 mm and 0.22±0.10 mm respectively. The radius of the meniscuses is 0.35±0.10 mm. These last three geometrical parameters, characterized by SEM (Scanning Electron Microscopy), show significant dispersions due to the elaboration process. SEM analysis has shown that walls of the HSPs contain many heterogeneities and defects such as porosities (fig. 1b).

a)
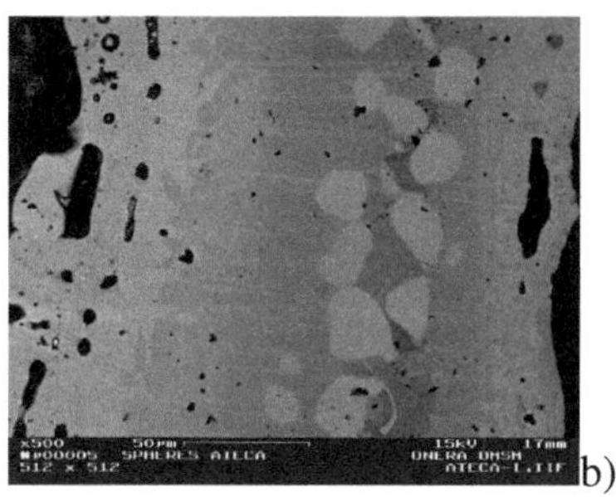
b)

Fig. 1 a) cylindrical compression sample b) SEM observation of microstructural heterogeneities in a meniscus

Uniaxial compression tests have been performed on a tension-compression machine at room temperature. Loads have been applied between two compression plates along the revolution axis of the cylindrical samples. Whereas the lower plate is fixed, experiments have been made for different imposed upper plate displacement rates of -1 mm/min, -10 mm/min and -100 mm/min, towards a compaction of ~14 mm.

2.2 *Experimental Results and Dispersion*

A first set of compression tests has been realized on five samples extracted from both the side and the core of the plate provided by Ateca at an upper plate displacement rate of -1 mm/min. Force vs. displacement curves corresponding to these experiments are shown on fig. 2a. These curves present many similarities with the ones classically observed for cellular materials like metal foams [3], *i.e.*, the shape of the curves reveals the existence of a densification plateau which is associated with the local plastic collapse of constitutive cells (hollow spheres in the case of HSPs). However, a second damage mechanism has been observed due to the local rupture of meniscuses in presence of defective brazes, thus samples tend to crumble during the compression tests. For this reason, results are plotted in term of force vs. displacement rather than stress vs. strain. Nevertheless, when reference is made to stress or strain in this paper concerning experimental results, they are computed from the initial diameter and length of compression samples.

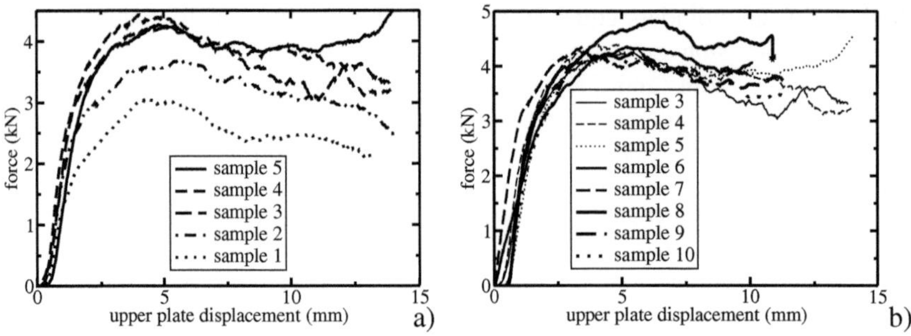

Fig. 2 a) boundary effect on the mechanical behaviour of HSPs due to their place of origin b) influence of the solicitation rate on the mechanical response of HSPs

It is worth noting that results dispersion is weak for samples extracted from the plate core (samples 3, 4 and 5) whereas the stiffness of samples vanishes drastically for samples cut near the side of the plate (samples 2 and 1). This effect has been verified performing compression tests on 15 other samples extracted from different places in the original plate provided by Ateca. Those results, not shown here, reveal that in a band of 60 mm (2 time the sample size approximately) surrounding the plate the compactness of the packing is disturbed so that the sample stiffness decreases. Such a boundary effect, in agreement with the works on spheres packings [1,8,14,15], could have a significant influence on small size packings and could be a relevant issue for many applications of such materials.

2.3 Influence of the Solicitation Rate

In order to avoid the boundary effect described previously, samples used for compression tests at different upper plate displacement rates have been cut in the core of the plate. Since the results dispersion on compression samples extracted from the core of the plate is weak, only 2 and 3 experiments have been performed at upper plate displacement rates of -10 mm/min (samples 6 and 7) and -100 mm/min (samples 8, 9 and 10) respectively. These results are compared with the ones obtained at -1 mm/min for samples 3, 4 and 5. Force vs. displacement curves are shown on fig. 2b. At room temperature any significant effect of the solicitation rate is observed, thus viscosity can be neglected for the material behaviour. Indeed, the mean plateau stresses (calculated as the mean maximum forces divided by the section of cylinders ~ 706.86 mm^2) equal 6.22 MPa, 6.17 MPa and 6.24 MPa at -1 mm/min, -10 mm/min and -100 mm/min respectively.

The aim of the following section is the development of a FE-modelling, based on the experimental characterization described previously, that would be able to appreciate and to predict accurately the compression behaviour of HSPs.

3 Finite-Element Modelling of Hollow-Sphere Packings Behaviour

3.1 Model Assumptions

The main damage mechanism that occurs during compaction tests is the plastic collapse of hollow spheres thus the rupture of meniscuses is neglected in the FE model. As one can expect, the architecture of HSPs induces stress concentration phenomena in the meniscuses which represent zones where a change in the geometry curvature occurs. Consequently, meshes are refined in these areas (see fig. 3a). Furthermore this area of the HSP architecture appears like a transition area between a volume component of the architecture, the meniscus, and a shell one, the hollow sphere. In this paper, the issue of the internal packing contact observed experimentally is not dealt with and calculations are stopped just before it appears.

In agreement with experimental characterization, the constitutive material is assumed to be elasto-plastic at the room temperature. Elasticity is linear isotropic and plasticity is modelled through an isotropic hardening using a von Mises criterion for the yield surface. The different moduli of the constitutive material behaviour are equal to E_m=200,000 MPa for the Young modulus, v_m=0.31 for the poisson's ratio, R_{0m}=60 MPa for the yield stress and H_m=10,000 MPa for the plastic modulus. The constitutive material is assumed to be near a pure Ni and to be homogeneous in the meniscuses and in the hollow spheres.

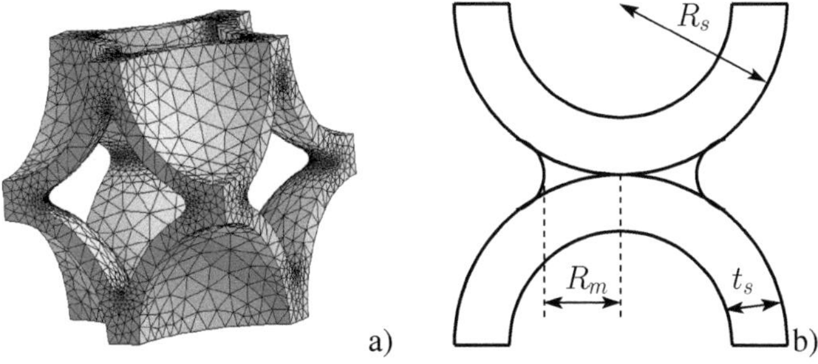

Fig. 3 a) unit cell of an infinite SC-like HSP (t_s/R_s=0.14 and R_m/R_s=0.35) b) geometrical parameters of HSPs

For the sake of simplicity, a regular Simple Cubic (SC)-like packing is considered. In this work, infinite SC-like packings are considered applying symmetry boundary conditions to the unit cell. Meshes are obtained with *BLSURF* [7] using quadratic tetrahedrons. The three different geometrical parameters that characterize the packings are: the outer radius of hollow spheres R_s, their thickness t_s and the radius of meniscuses R_m (see fig. 3b). The issue of the dispersion experimentally observed on these parameters is not tackled with in this work. A displacement is imposed on the top of the unit cell with a rate of -0.6 mm/min, close to the one applied experimentally (despite it is not a critical issue cause viscosity is neglected in this study). Calculations are performed using the FE-code *ZSet* [16,17].

3.2 Calculation Predictions

In order to characterize the roles of the constitutive parts of HSPs (hollow spheres on one side and meniscuses on the other side) on their effective mechanical behaviour, 25 different geometries have been simulated. Five values for both ratios t_s/R_s and R_m/R_s have been chosen ranging between 0.10 and 0.26 with an increment of 0.04 and between 0.20 and 0.40 with an increment of 0.05 respectively. For compression samples those ratios are close to 0.14 and 0.25 respectively.

As expected, cumulated plastic strain maps show that plasticity occurs mainly in the meniscuses and in the transition area between meniscuses and hollow spheres (see fig. 4a). Von Mises stress maps, not shown here, assert these results. Stress vs. strain curves are plotted on fig. 5, only for some of the 25 geometries simulated for sake of clarity. ε_{33} and σ_{33} denote the components of respectively the strain and the stress tensors along the solicitation direction. Stress vs. strain curves show that internal-packing contact occurrence is strongly dependent on the meniscus radius (calculations have been stopped just before its occurrence), see fig. 5a. Indeed, the ends of the meniscuses appears to work like plastic articulations around which the collapse of hollow spheres occurs (see fig. 4b), thus, the bigger the meniscuses, the farther are these articulation points are and the more the occurrence of internal packing contact is dismissed for higher global strain levels. If these articulation points are far enough, a decrease of the packing stiffness can be observed associated with the buckling of hollow spheres which induces a double curvature of hollow spheres in this area (see fig. 4b). That last phenomenon is all the more significant since hollow sphere thickness decreases. Nevertheless, the strain level for which the internal-packing contact occurs does not seem strongly dependent on the ratio t_s/R_s.

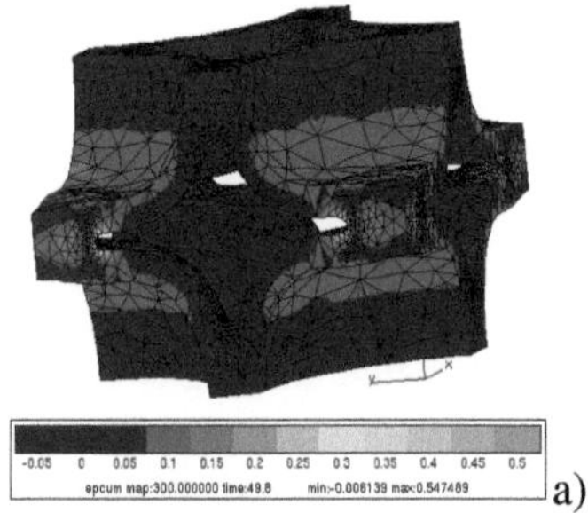 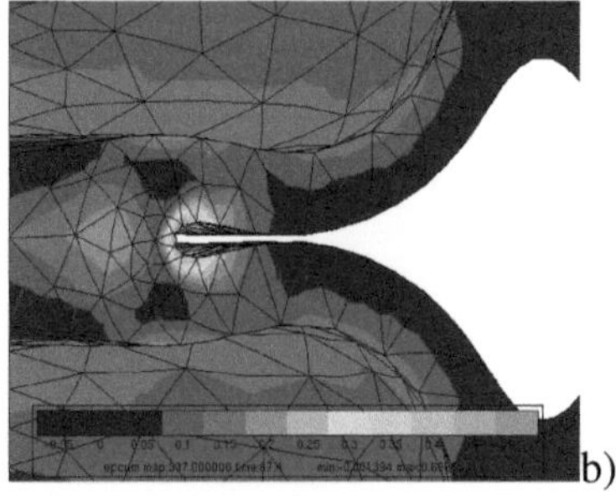

Fig. 4 a) cumulated plastic strain map for a global strain of 25 % b) double curvature induced by the buckling of hollow spheres for high global strain levels, here 34 % ($t_s/R_s=0.14$ and $R_m/R_s=0.35$)

Effective behaviour of HSPs has been determined by fitting stress vs. strain curves between 0 % and 1 % to the mechanical response of a Homogeneous Equivalent Medium. In view of the shape of stress vs. strain curves, a non linear exponential term has been added to the hardening in order to fit correctly their curvature; a fit of the effective behaviour of HSPs with the same behaviour as their constitutive material was no relevant. Indeed the geometry of the HSPs induces a non linearity due to the heterogeneity of stress distribution. Normalized effective Young modulus E^{eff}/E_m and yield stress R_0^{eff}/R_{0m} are plotted as function of the packing volume fraction on fig. 6a (calculated as the sum of the tetrahedrons meshing a given geometry divided by $8R_s^2$ the volume of the unit cell). The stiffness of HSPs increases with the hollow sphere thickness and, in a more significant manner, with the meniscus radius. It is worth noting that the effective stiffness of the packing results in a combination between the stiffness of the hollow spheres and the one of the meniscuses (through the ratios t_s/R_s and R_m/R_s respectively). Nevertheless, the contributions of both constitutive parts of HSPs seem to be relatively independent.

4 Comparison Between Calculation Predictions and Experimental Results

FE calculations have shown the model developed is quite accurate to appreciate the different mechanisms that occur during the quasi-static compression of HSPs. That qualitatively good agreement between modelling and experiment is valid for small compaction levels but also for higher ones. Results obtained highlight the important role of non-linear phenomena in the compression behaviour of such materials. Moreover, the effective stiffness of HSPs seems to be strongly dependant on the respective stiffness of hollow spheres and meniscuses. However, as one could be expect, the model does not seem to be able to appreciate the localized collapse in bands of the constitutive cells usually observed in experimental studies. Indeed, the proposed model reduces the HSP to its unit cell, consequently the distribution of the plasticity is assumed to be uniform in all the HSP and the issue of the localization in a particular band of cells cannot be raised.

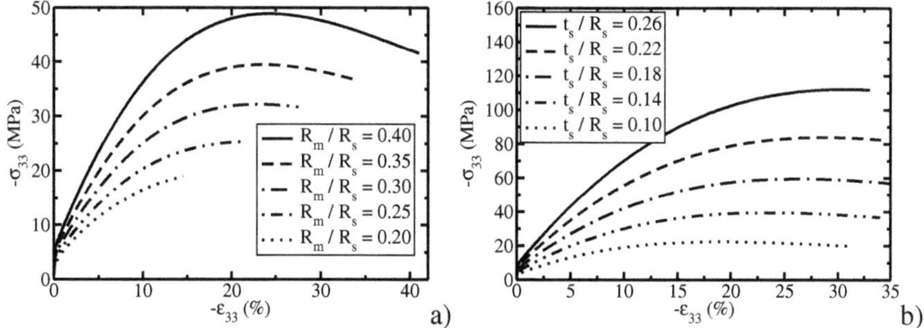

Fig. 5 influence of the respective stiffness of hollow spheres and meniscuses on the effective stiffness of HSP a) influence of the ratio R_m/R_s, with t_s/R_s=0.14 b) influence of the ratio t_s/R_s, with R_m/R_s=0.35

On the contrary to the qualitative comparison between modelling and experiment, the quantitative comparison is not so conclusive. Indeed, calculations strongly overestimate the packings stiffness (see fig. 6b). Such a disagreement could be explained by an incorrect choice of the constitutive material behaviour. In fact, as mentioned in the modelling assumptions, the constitutive material has been supposed to be near a pure bulk Ni. However SEM micrographs have revealed that the microstructure of the constitutive material is strongly heterogeneous (see fig. 1b). Then, for so thin architectures, a second scale of heterogeneity appears, associated to the constitutive material that could not be considered as a bulk material.

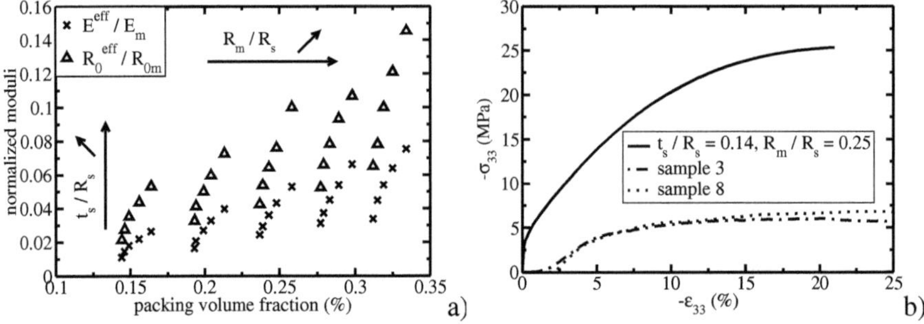

Fig. 6 a) normalized effective young modulus and yield stress as function of the packing volume fraction b) comparison between FE-modelling predictions and experiments (curves for samples 3 and 8 on fig. 2b have been converted in stress vs. strain ones)

5 Conclusions

Compression behaviour of HSPs shows similarities with the one of metal foams. Experimentally, a densification plateau is observed associated with the local collapse of hollow spheres. For small strains the behaviour of HSPs is mainly governed by the local plasticity in the meniscuses neighbourhood, whereas for higher strain levels buckling of hollow spheres and internal-packing contact occur.

A detailed study of regular HSPs has been proposed through FE-modelling. Calculations have shown a qualitatively good agreement with experimental results. Mechanisms which occur during compaction of such materials have been correctly appreciated by the model even for high strain levels. Calculations revealed that the overall stiffness of such materials is strongly influenced by the stiffness of the hollow spheres and the meniscuses in a relatively independent way. However, a quantitative comparison between modelling predictions and experiments has shown that calculations strongly overestimate the mechanical behaviour of HSPs and raises the issue of the choice of the behaviour of their constitutive material that could be no longer considered as a bulk one. This issue is the topic of complementary works on HSPs.

Acknowledgments. This work was performed within the *'Aerodynamic and Thermal Load Interactions with Lightweight Advanced Materials for High Speed Flight'* project investigating high-speed transport. ATLLAS, coordinated by ESA-ESTEC, is supported by the EU within the 6th Framework Programme Priority 1.4, Aeronautic and Space. The firm Ateca is gratefully acknowledged for providing non-regular hollow-sphere packings.

References

[1] Aste, T., Saadaftar, M., Senden, T.J.: Geometrical structure of disordered sphere packings. Physical Review E 71(6), 61302 (2005)

[2] Brothers, A.H., Dunand, D.C.: Plasticity and damage in cellular amorphous metals. Acta Mat. 53, 4427–4440 (2005)

[3] Evans, A.G., Hutchinson, J.W., Ashby, M.F.: Multifunctionality of cellular metal systems. Harvard University, Cambridge (1997)

[4] Gibson, L.J., Ashby, M.F.: Proc. R. Soc. of London, A 382, 43 (1982)

[5] Gibson, L.J., Ashby, M.F.: Cellular solids: structure and properties, 2nd edn. Cambridge University Press, Cambridge (1997)

[6] Hodge, A.M., Dunand, D.C.: Measurement and modeling of creep in open-cell NiAl foams. Metallurgical and Materials Transactions A 34, 2353–2363 (2003)

[7] Laug, P., Borouchaki, H.: BLSURF - mailleur de surfaces composées de carreaux paramétrés - manuel d'utilisation. BLSURF user manuel, INRIA repport no. 0232 (1999)

[8] Müeller, G.E.: Numerically packing spheres in cylinders. Powder Technology 159, 105–110 (2005)

[9] Paul, A., Ramamurty, U.: Strain rate sensitivity of a closed-cell aluminium foam. Mat. Sci. Engng. A 281, 1–7 (2000)

[10] Rakow, J.F., Waas, A.M.: Size effects and the shear response of aluminium foam. Mechanics of Materials 37, 69–82 (2005)

[11] Ruan, D., Lu, G., Chen, F.L., Siores, E.: Compressive behaviour of aluminium foams at low and medium strain rates. Composite Structures 57, 331–336 (2002)

[12] Sanders, W.S., Gibson, L.J.: Mechanics of hollow sphere foams. Mat. Sci. Engng., A 347, 70–85 (2003)

[13] Sanders, W.S., Gibson, L.J.: Mechanics of BCC and FCC hollow- sphere foams. Mat. Sci. Engng., A 352, 150–161 (2003)

[14] Scott, G.D.: Packing of equal spheres. Nature 188, 908–909 (1960)

[15] Scott, G.D., Kilgour, D.M.: The density of random close packing of spheres. British Journal of Applied Physics 2(2), 863–866 (1969)

[16] ZeBuLoN guide for material behaviour laws, version 8.3.6 (2006)

[17] ZeBuLoN user manuel, version 8.3.6 (2006)

Modelling the Smooth Muscle Tissue as a Dissipative Microstructured Material

Fanny Moravec and Miroslav Holeček

Abstract. A highly simplified model of the smooth muscle tissue is formed to study a special form of viscous effects occurring in living tissues. Namely, when a sample of the tissue is deformed, the fluid in the extracellular matrix has to move because it is extruded from one places into others. This phenomenon contributes into the global viscous properties of the tissue. That contribution is studied by means of a simplified model of smooth muscle tissue regarded as a regular lattice of elastic inclusions. An expression of the dissipation due to the fluid flow around the inclusions is found out. At the macroscopic (continuum) level, the idea of constitutive model with internal variables is used. As a result, a differential form of the viscoelastic behaviour is derived. The model is solved numerically and fitted by a confrontation with experiment performed on gastropod smooth muscle tissue.

1 Introduction

The smooth muscle tissue is a very complex material with a sophisticatedly organized structure from the molecular to macroscopic level. It is worth studying its mechanical properties to understand the way in which the tissue is able to present an optimal mechanical behaviour. It is however very difficult to recognize perfectly the mechanical properties of even one living cell [1]. A detail insight into mechanics of the whole tissue forming by many interacting and cooperating cells seems to be a hopeless task. Nevertheless, some chosen features of its mechanical behaviour may be studied by *idealized models* that highlight some aspects and neglect others.

Fanny Moravec
Department of Mechanics, University of West Bohemia in Pilsen,
Univerzitni 22, 306 14 Pilsen, Czech Republic
e-mail: `fanny@kme.zcu.cz`

Miroslav Holeček
Department of Mechanics, University of West Bohemia in Pilsen,
Univerzitni 22, 306 14 Pilsen, Czech Republic
e-mail: `holecek@kme.zcu.cz`

J.-F. Ganghoffer, F. Pastrone (Eds.): Mech. of Microstru. Solids, LNACM 46, pp. 101–108.
springerlink.com

An example of such a highly idealized structure is as follows. It consists of a regular lattice of elastic inclusions that represent individual cells of the tissue. The neighbouring inclusions are coupled by linear elastic forces that correspond to mechanical properties of the extracellular matrix at small strains. Elastic forces related to the strain of inclusions themselves are also linearized. The whole structure, however, is nonlinear because of the constrain corresponding to the incompressibility of individual inclusions. This model has been used in previous work [3] to study the remarkable ability of the smooth muscle tissue to tune its global stiffness by changes at the level of cellular cytoskeleton (a biopolymer network spanning each eukaryotic cell).

In this work, we use a modification of this model to study the *viscoelastic* properties of these tissues. As well known, polymer networks has both elastic and viscous behaviour. The viscoelastic response under dynamic excitation of biopolymer networks is studied in many works (see e.g. [7]). In the whole tissue, however, there is also another phenomenon contributing to its global viscous properties. Namely, when straining the tissue, the fluid filling the space between cells has to move. The reason is that the gap between neighbouring cells slightly varies and the fluid has to fill or drain the spaces. This movement is connected with some viscous forces that may be introduced in our model. The resulting continuum limit describes a material with *internal variables* [4]. That is why we present briefly the use of internal variables in continuum modelling in the next section. In the third section, these internal variables are identified with special geometric (microscopic) parameters of our model which allows us to formulate the complete set of evolution equations at the macroscopic scale. In the fourth section, these equations are numerically solved in the case of relaxation test and the results are compared with experimental data.

2 Constitutive Model with Internal Variables

In continuum description, the thermodynamic state of elastic materials is fully determined by two variables distributed in space, $\mathbf{F}$ and Θ, the gradient of the deformation and the temperature distribution, respectively. Concerning dissipative materials, other state variables are necessary to characterize the inelastic behaviour of the material. Such variables usually cannot be defined as some macroscopic quantities. Nevertheless, we suppose a set of some "hidden" ones that are not directly accessible by experimental measurement at the macroscopic level. We call them the *internal variables*.

Let us suppose an occurrence of n scalar internal variables, β_i, $i = 1,\ldots,n$. If all processes are isothermal, an explicit dependence on the temperature may be omitted. The fact that the internal variables are state parameters implies that the Helmholtz free energy function may be written in the form $\Phi = \Phi(\mathbf{F},\boldsymbol{\beta})$ (where $\boldsymbol{\beta}$ represents the set of all internal variables β_i). Following [4], we assume the dissipation in the material as characterized by functions Ξ_i such that the internal dissipation, D_{int}, is described as

$$D_{int} = \sum_{i=1}^{n} \Xi_i(\mathbf{F}, \dot{\mathbf{F}}, \boldsymbol{\beta}, \dot{\boldsymbol{\beta}})\, \dot{\beta}_i, \tag{1}$$

where the dot denotes the time derivative of the state variables. The second law of thermodynamics claims the validity of the Clausius-Planck inequality, namely

$$D_{int} = w_{int} - \dot{\Phi} \geq 0, \tag{2}$$

where $w_{int} \equiv \Pi : \dot{\mathbf{F}}$ is the power of internal mechanical work and Π is the first Kirchhoff stress measure. Applying the chain rule for the time derivative of the free energy, we obtain from (2) that

$$D_{int} = \left(\Pi - \frac{\partial \Phi}{\partial \mathbf{F}}\right) : \dot{\mathbf{F}} - \sum_{i=1}^{n} \frac{\partial \Phi}{\partial \beta_i} \dot{\beta}_i \geq 0. \tag{3}$$

To fulfill the inequality (3) in all admissible processes the constitutive law,

$$\Pi = \frac{\partial \Phi(\mathbf{F}, \boldsymbol{\beta})}{\partial \mathbf{F}}, \tag{4}$$

has to be valid. With (1), it implies that the time evolution of the n internal variables β_i is governed by n differential equations,

$$\Xi_i(\mathbf{F}, \dot{\mathbf{F}}, \boldsymbol{\beta}, \dot{\boldsymbol{\beta}}) + \frac{\partial \Phi(\mathbf{F}, \boldsymbol{\beta})}{\partial \beta_i} = 0. \tag{5}$$

Finally, the modelling of dissipative material with internal variables reduces into these questions: which are the internal variables β_i and which expressions have to be chosen for the functions Φ and Ξ_i?

Considering the viscoelastic behaviour, the free energy function Φ should result from the elastic forces while the dissipative functions Ξ_i should depend on the viscosity of the material. This separation between elastic and viscous characters is allowed by the choice of the internal variables, β_i, as suitable microscopic parameters. For instance, the well known one-dimensional generalized Maxwell model (ie. a certain number of parallel branches formed by a spring and a dashpot in series, in parallel with a branch formed by an alone spring) works with the deformation of dashpots as internal variables, see eg. [4]. In this model, the free energy Φ corresponds to the total strain energy of springs while the functions Ξ_i reduce to the products of the rates of the internal variables, $\dot{\beta}_i$, and the constants of viscosity of the corresponding dashpot, η_i, ie. $\Xi_i = \eta_i \dot{\beta}_i$. Consequently, the differential equations (5) are mutually independent.

3   The Idealized Model of Smooth Muscle Tissue

Consider an idealized material which microstructure is formed by anisotropic inclusions regularly arranged in a homogeneous matrix. Suppose that the *representative volume element* (RVE) may be defined as a rectangle including one inclusion and a suitable part of the surrounding matrix. We will study only such deformations of this material in which each RVE remains rectangular. It means that only the pure strain is studied and no shear deformation is supposed. The actual shape of a RVE is determined by the actual sizes of its edges, Δx_i, which are related to the reference ones, Δx_i^{ref}, via the definition of the *principal stretches*, $\lambda_i = \Delta x_i / \Delta x_i^{ref}$ ($i = 1,2,3$). At the microscopic scale, we simplify considerably the description by assuming that only *six* microscopic variables characterize the inner geometric configuration of any RVE. The first three ones, c_i, characterize dimensions of the inclusion in the directions of the rectangle edges. The last ones, β_i, measure distances between neighbouring inclusions in the same directions, see Fig. 1. When assuming that the inclusion is always in the center of a RVE, the three geometric relationships,

$$c_i + \beta_i = \Delta x_i, \quad i = 1,2,3, \tag{6}$$

have to be fulfilled at any time. Consequently, studying the mechanical behaviour of such material asks for the definition of *three* internal variables, say β_i, $i = 1,2,3$.

The viscoelastic properties of the material are described as follows. We assume the matrix to be filled with a compressible viscous substance, while the substance constituting the inclusion is assumed to be (quasi)incompressible and not viscous. Namely, when interpreting the structure as a simplified model of a living tissue, we may suppose that elastic properties originate from biopolymer networks distributed both in the matrix and within inclusions (the latter forming the cytoskeleton). We approximate that elastic properties by means of six effective stiffness coefficients, denoted K_i^β and K_i^c, $i = 1,2,3$, occuring in the matrix and in the inclusions respectively. For simplicity, we assume that the anisotropy of the problem has only a geometrical origin [3]. That is, we approximate the biopolymer networks as the regular repetitions of representative elements (springs). We introduce the relative stiffness k as the ratio of the stiffness coefficients of the representative elements in inclusion and in matrix. Thus, we can relate the effective stiffness to a chosen one, say K_1^β, by the relationships

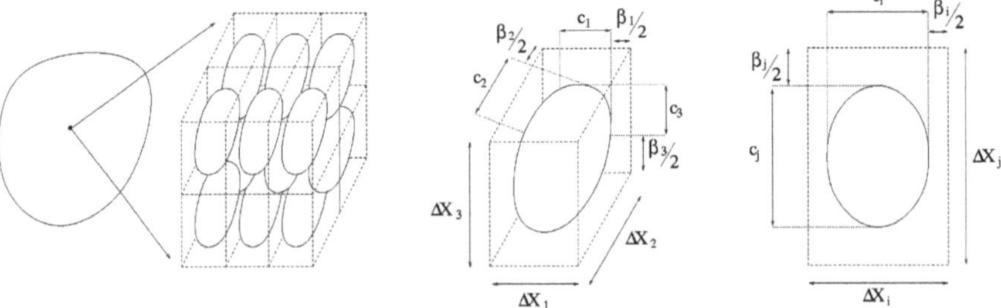

Fig. 1 Schematic microstructure of the studied periodic material

$$K_i^c = k \, \frac{c_1^{(0)} \beta_1^{(0)}}{c_i^{(0)2}} \, K_1^\beta, \ \text{and} \ K_i^\beta = \frac{c_1^{(0)} \beta_1^{(0)}}{c_i^{(0)} \beta_i^{(0)}} \, K_1^\beta, \ i = 1,2,3, \tag{7}$$

where the up index $^{(0)}$ denotes the rest lengths of the effective elastic springs representing the linearized elastic properties of the matrix and the inclusions. For simplicity, we consider that the ratio between rest lengths in different directions is the same for the inclusion and for the matrix, ie. we assume $c_i^{(0)}/c_j^{(0)} = \beta_i^{(0)}/\beta_j^{(0)}$. If moreover the material is considered as transverse isotropic (ie. the properties are the same at two directions but different from the third one), then only one parameter, namely $r \equiv c_1^{(0)}/V_c^{1/3}$ where $V_c \equiv \prod_{i=1}^{3} c_i$ is related to volume of inclusion, suffices to fully define the material anisotropy. Namely, if $r = 1$ the material is isotropic. If $r > 1$ the inclusion and the RVE are long shaped. If $r < 1$ their are flattened.

The use of linearized elastic forces between individual elements at the microlevel (at RVE) leads to a simple form of the Helmholtz free energy, Φ,

$$\Phi = \frac{1}{2V_{rve}} \sum_{i=1}^{3} \left(K_i^c \left(c_i - c_i^{(0)} \right)^2 + K_i^\beta \left(\beta_i - \beta_i^{(0)} \right)^2 \right) + \frac{1}{\varepsilon} \left(\frac{V_c}{V_c^{(0)}} - 1 \right)^2, \tag{8}$$

where the limit $\varepsilon \to 0$ expresses the incompressibility of the inclusions (in fact, ε is chosen as a very small number in numerical modelling). Namely $V_c^{(0)} \equiv \prod c_i^{(0)}$ is related to the volume of the inclusion at the rest configuration. Similarly $V_{rve} \equiv \prod_{i=1}^{3} \Delta x_i^{ref}$ is the reference volume of each RVE. For convenience, we define the relative volume ratio $v \equiv V_c^{(0)}/V_{rve}$. Note that $0 < v < 1$ in any case.

The internal dissipation, D_{int}, is derived from the power of viscous forces in the matrix. The viscous forces in our model are generated *only* by the movement of the fluid in the matrix. At a time moment, the gap between neighbouring inclusions, say in the i-the direction, has the characteristic actual dimension β_i which increases (or decreases) with the velocity $\dot{\beta}_i$. The actual dissipative forces related to the flow filling (or draining) this gap, f_i, may be estimated to be dependent on β_i, $\dot{\beta}_i$, the viscosity η, the fluid density ρ and a characteristic surface, S_i, measuring the dimension of the gap in the direction perpendicular to the coordinate β_i. That is, we suppose $f_i = f_i(\beta_i, \dot{\beta}_i, S_i, \eta, \rho)$. When employing the dimensional analysis, the function f_i may be only of the form

$$f_i = \eta \beta_i \dot{\beta}_i g \left(\frac{S_i}{\beta_i^2}, Re \right), \tag{9}$$

where g is a function and Re is the Reynolds number of the problem. Let us compare this formula with an expression obtained by the solution of the Navier-Stokes equation in a concrete case. For example, A.Ya. Malkin [6] found that the force needed for the extrusion of a fluid with the viscosity η from the space defined by two parallel discs with the radius R is

$$f_i = \frac{\pi \eta \dot{h} R^4}{h^3} = \frac{\eta S^2 \dot{h}}{\pi h^3}, \tag{10}$$

where h is the current distance between the discs. Comparing with (9) we notice that an explicit dependence on the Reynolds number does not occur, and that g is a quadratic function. Thus, we propose to approximate the viscous forces by the form

$$f_i = \frac{\kappa \eta S_i^2 \dot{\beta}_i}{\beta_i^3}, \tag{11}$$

where κ is a (dimensionless) number. The characteristic surface S_i may be expressed by the product $c_j c_k$, $j \neq i \neq k \neq j$, which is proportional to the projection of the inclusion surface onto the ith direction. Defining the internal dissipation as the density of the sum of powers of viscous forces in the three directions, ie. by integrating the viscous forces f_i with respect to the velocity $\dot{\beta}_i$, we obtain

$$D_{int} = \frac{\eta \kappa}{2 V_{rve}} \sum_{i=1}^{3} \frac{(c_j c_k)^2}{\beta_i^3} \dot{\beta}_i^{\,2}, \quad j \neq k, \ i \neq j, \ i \neq k. \tag{12}$$

Using (1) we write the function Ξ_i as

$$\Xi_i(\mathbf{F}, \dot{\mathbf{F}}, \boldsymbol{\beta}, \dot{\boldsymbol{\beta}}) = \dot{\beta}_i \, \frac{\tau K_1^\beta (c_j c_k)^2}{V_{rve}^{4/3} \beta_i^3}, \tag{13}$$

where $i, j, k = 1, 2, 3$, $j \neq k$, $i \neq j$, $i \neq k$ and $\tau \equiv \eta \kappa / (2 K_1^\beta V_{rve}^{-1/3})$ is a characteristic time of the problem. Putting the expressions (13) and (8) into (5) we get the system of differential equations governing the time evolution of the internal variables β_i. It must be, however, solved numerically.

4   Numerical Simulation

For simplicity, numerical simulations are done on a material macroscopically incompressible, transverse isotropic, loaded unidirectionally along the direction of anisotropy, say $i = 1$. Thus

$$\mathbf{F} = \begin{pmatrix} \lambda & 0 & 0 \\ 0 & \lambda^{-1/2} & 0 \\ 0 & 0 & \lambda^{-1/2} \end{pmatrix}. \tag{14}$$

For macroscopically incompressible materials the constitutive law (4) has to be rewritten to include a macroscopic hydrostatic pressure, p, ie.

$$\Pi = \frac{\partial \Phi}{\partial \mathbf{F}} - p \frac{\partial \det F}{\partial \mathbf{F}}. \tag{15}$$

Using the fact that stress vanishes in normal directions to the direction of traction, $\Pi_{22} = \Pi_{33} = 0$, an expression for p can be found and finally the material behaviour law reduces to the law

$$\Pi_{11} = \frac{\partial \Phi(\lambda, \beta_i)}{\partial \lambda}, \tag{16}$$

associated with the differential equations (5). It is worth stressing that in equation (16) the variables λ and β_i have to be considered as *independent* and that the derivation is done with respect to λ for β_i fixed.

The material behaviour was simulated during the stress relaxation test. The deformation of the sample is controlled through the time evolution of the stretch λ, which time evolution is composed of two parts. Firstly, the stretch quickly increases to reach a given value, say λ_{cst}. In this part, its time evolution is approximated using the third order polynomial to make the function $\lambda(t)$, as well as its rate $\dot{\lambda}(t)$, smoother. Secondly, during a long period of time, the strain is maintained constant by keeping the stretch at the value λ_{cst}. The differential equations (5) are solved numerically using the function ODE23 of Matlab environment (The MathWorks, Inc., Mattick, Massachusetts). The initial solution is chosen assuming the initially undeformed material to be stress free, i.e. $\beta_i|_{t=0} = \beta_i^{(0)}$, $c_i|_{t=0} = c_i^{(0)}$ and $\dot{\beta}_i|_{t=0} = 0$, recalling that the up index (0) is used to denote the elastic fibres rest lengths. As a result, we plot in Fig.2 the stress evolution $\sigma(t')$ when the strain is kept constant, t' being the time t minus the time necessary to reach the value of strain λ_{cst}. The material behaviour is qualitatively influenced by the time τ and the three dimensionless parameters k, r, and v. The two dimensional parameters, ie. the rigidity K_1^{β} and the volume V_{rve}, do not influence it qualitatively but the stress is proportional to the quantity $K_1^{\beta}/V_{rve}^{1/3}$.

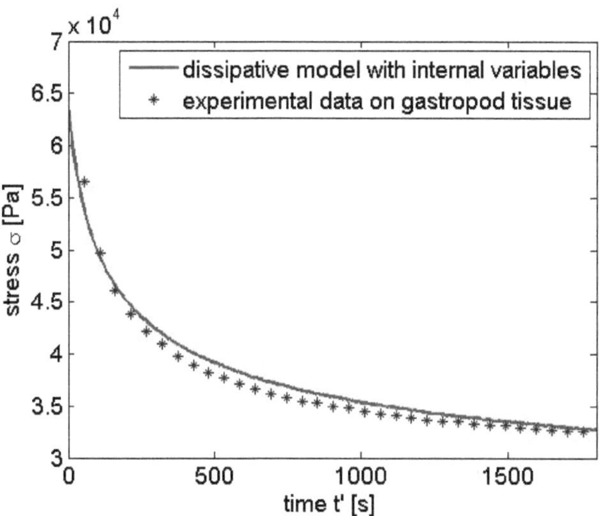

Fig. 2 Confrontation between the stress relaxation of 187% strained gastropod tissue and the dissipative model with parameters $r = 3$, $v = 0.6$, $k = 100$, $\tau = 300$s, $K_1^{\beta}/V_{rve}^{1/3} = 450$Pa, $\varepsilon = 0.001$

The result of the simulation is confronted with the experimental data obtained from [5]. Measurements (relaxation tests) on unfixed samples of smooth muscle extracted from the soles of gastropods foots were obtained using the apparatus DMA 7 (dynamic mechanical analyzer, Perkin Elmer INC, Wellesley, Massachusetts). The large strain of 87% (ie. $\lambda_{cst} = 1.87$) was applied to the samples. Model parameters are identified from the Fig.2 and show that, in best fitted model, the muscular cells of gastropod tissue are approximated as long ($r = 3$) and very stiff ($k = 100 >> 1$) inclusions. Such a stiffness may be hardly realistic. Nevertheless, we cannot forget that our model describe only a part of the global viscoelastic behaviour. The effects related to own viscosity of polymer networks has not be taken into account.

5 Conclusion

We have modified the previously formed model of the smooth muscle tissue [3] with the view to describe a special viscous effect in the tissue. The effect is connected with a movement of the extracellular fluid during deformations of the tissue. We have found the continuum limit of this model in which that viscous phenomena is described by means of internal variables. These variables has a clear geometric origin because they measure an actual gap between neighbouring cells. Our results show that the influence of the viscosity of the extracellular fluid cannot be neglected in dynamics of a macroscopic tissue sample. For instance, the stress relaxation phenomenon is well observed for this material. The perspective of this contribution consists also in an apparent transparency of the used materials parameters. The identification of model parameters via experiments with macroscopic samples thus may give a partial information about tissue microstructure. Moreover, an influence of material microstructure on the macroscopic mechanical behaviour can be studied. The model thus could be also helpful for quantifying the effect of cellular pathologies on the global tissue behaviour and organs functioning.

Acknowledgements. This work has been supported by the Czech Ministry of Education, Sports and Youth, Project No. MSM4977751303.

References

1. Boal, D.: Mechanics of the Cell. Cambridge University Press, Cambridge (2003)
2. Fung, Y.C.: Biomechanics: Mechanical Properties of living Tissues. Springer, Heidelberg (1993)
3. Holeček, M., Moravec, F.: Hyperelastic model of a material which microstructure is formed by "balls and springs". Int. J. Solids and Struct. 43, 7393–7406 (2006)
4. Holzapfel, G.A.: Nonlinear Solid Mechanics. John Wiley & Sons, LTD, Chichester (2001)
5. Kochová, P., Tonar, Z.: 3D reconstruction and mechanical properties of connective and smooth muscle tissue. In: proceedings of Human Biomechanics, Congress of the Czech Society of Biomechanics, pp. 1–9 (2006)
6. Malkin, A.Y.: Rheology Fundamentals. ChemTec Publishing (1994)
7. Storm, C., Pastore, J.J., MacKintosh, F.C., Lubensky, T.C., Janmey, A.P.: Nonlinear elasticity in biological gels, gels. Nature 435, 191–194 (2005)

Mechanical Response of Helically Wound Fiber-Reinforced Incompressible Non-linearly Elastic Pipes

Paola Nardinocchi, Tomas Svaton, and Luciano Teresi

Abstract. We study the mechanical response of a helically wound fiber-reinforced incompressible axisymmetric structure under torsion and compare it with the response turning out from the classical Rivlin solution of the torsion problem of a neo-Hookean pipe.

1 Introduction

In the last few years, fiber-reinforced materials have made a new comeback due to the bio-mechanical involvement. Indeed, a lot of biological tissues are characterized by a material response that is strongly anisotropic due to a fiber structure which determines the relevant mechanical behavior of the tissue and the problem of characterizing the behavior of fiber-reinforced media undergoing finite deformations turns out as a relevant theme in literature [1], [2]. The macroscopic description of the material response of fiber-reinforced materials is typically given in terms of a strain-energy function which is dependent on specific deformation invariants [8]. Moreover, it is usually assumed that such function be additively decomposable into an isotropic part associated with the isotropic base material and an anisotropic part accounting for the anisotropic character of the material due to the reinforcement. Recently, a lot of analysis has been devoted to show as different phenomena related to fiber-reinforced materials may be captured within this framework [3], [4], [5].

Paola Nardinocchi
Dipartimento di Ingegneria Strutturale e Geotecnica, Università di Roma "La Sapienza"
via Eudossiana 18, I-00147 Roma, Italy
e-mail: `paola.nardinocchi@uniroma1.it`

Tomas Svaton
University of West Bohemia, Pilsen, Czech Republic
e-mail: `tsvaton@gmail.com`

Luciano Teresi
Dipartimento di Strutture, Università Roma Tre, Roma, Italy
e-mail: `teresi@uniroma3.it`

J.-F. Ganghoffer, F. Pastrone (Eds.): Mech. of Microstru. Solids, LNACM 46, pp. 109–117.
springerlink.com   © Springer-Verlag Berlin Heidelberg 2009

Here, we analyze the mechanical response of a helically wound fiber-reinforced incompressible cylindrical structure to torsional deformations. The pure torsional deformation has been studied for isotropic and incompressible cylinders by Rivlin [7]. We extend the Rivlin's solution to helically wound fiber-reinforced incompressible cylinders and cylindrical pipes and study the rich mechanical behaviour due to the anisotropy induced by the fibers. From our point of view, this is a first step towards the analysis of the torsional deformation induced during the cardiac cycle in the left ventricles which may be roughly viewed as an ellipsoid of revolution consisting of a non–linearly elastic and incompressible material with embedded helicoidal muscle fibers whose angle changes linearly across the wall (recent studies have shown that the left ventricular torsion is an important indicator of cardiac function).

2 Fiber-Reinforced Bodies and Torsional Deformations

In 1940s, R.S. Rivlin wrote a short series of papers on the large elastic deformations of isotropic and incompressible materials with special attention to some simple problems: flexure, shear, and torsion of cylinder-like axisymmetric bodies. Here, the Rivlin solution of the pure torsion problem of elastic, incompressible and cylinder-like bodies is extended to helically wound fiber-reinforced incompressible elastic pipes and cylinders. The classical Rivlin solution is recovered as a special case.

Let $\{\mathbf{e}_1, \mathbf{e}_2, \mathbf{e}_3\}$ be an orthonormal basis of the vector space $\mathcal{V} = T\mathcal{E}$, with $\mathcal{E}$ the three-dimensional Euclidean space, and let $\mathcal{U}$ be the orthogonal complement to $\mathrm{span}\{\mathbf{e}_3\}$ with respect to $\mathcal{V}$. In a plane $\mathcal{P} \perp \mathbf{e}_3$ let us consider the connected and compact domain $\mathcal{D} = \mathcal{D}_o \setminus \Lambda_o$ with boundary $\partial\mathcal{D} = \partial\mathcal{D}_o \bigcup \partial\Lambda_o$ where $\mathcal{D}_o$ and the lacuna Λ_o, if present, are simply connected regions. Fixed an interval $\mathcal{I} = [0, h] \subset \mathcal{R}$, we consider the axisymmetric cylinder-like region $\mathcal{C} = \mathcal{D} \times \mathcal{I}$; $\mathcal{D} \times \{0\}$ and $\mathcal{D} \times \{h\}$ are the bases of $\mathcal{C}$, and $\partial\mathcal{D} \times \mathcal{I}$ is the mantle of $\mathcal{C}$, eventually given by the inner mantle $\partial\Lambda_o \times \mathcal{I}$ and the outer mantle $\partial\mathcal{D}_o \times \mathcal{I}$. Fixed a pole $o \in \mathcal{P}$, the place $\mathbf{p}$ of any $p \in \mathcal{C}$ with respect to o is given by the vector field

$$\mathbf{p} = r\,\mathbf{n}(\theta) + \zeta\,\mathbf{e}_3\,, \quad \mathbf{n}(\theta) = \cos\theta\,\mathbf{e}_1 + \sin\theta\,\mathbf{e}_2,\ 0 < \theta < 2\pi,\ \zeta \in \mathcal{I};\quad (1)$$

moreover, $0 < r < R_o$ when $\mathcal{D} \equiv \mathcal{D}_o$, $R_i < r < R_o$ otherwise[1]. A helix-like fiber of pitch b is a curve whose unit tangent vector $\mathbf{e} = \hat{\mathbf{e}}(r, \theta)$ at any $p \in \mathcal{C}$ is defined as

$$\hat{\mathbf{e}}(r, \theta) = \frac{r}{(r^2 + b^2)^{\frac{1}{2}}}\,\mathbf{n}_{,\theta}(\theta) + \frac{b}{(r^2 + b^2)^{\frac{1}{2}}}\,\mathbf{e}_3\,. \quad (2)$$

So, at any place $\mathbf{p}$ of a helically wound fiber-reinforced body $\mathcal{C}$, the material fiber $(\mathbf{p}, \mathbf{e})$ with $\mathbf{e}$ as in (2) identifies a locus of material reinforcement; for $b = 0$

---

[1] With this, an element $p \in \mathcal{C}$ is equivalently identified by the place $\mathbf{p}$ and by the cylindrical coordinates (r, θ, ζ).

equation (2) defines a circumferential uniaxial reinforcement. An interesting case is represented by a helically wound fiber-reinforced cylindrical pipe with pitch $b = b(r)$.

The Pure Torsional Deformation: Strain and Stress

In [7], Rivlin introduces the torsional deformation $\mathscr{C} \ni p \mapsto x = \mathbf{f}(p) \in \mathscr{E}$ of the cylinder-like body in which planes orthogonal to $\mathrm{span}\{\mathbf{e}_3\}$ remain plane and suffer only a pure rotation about $\mathbf{e}_3$. A concise representation formula for $\mathbf{f}$ is the following

$$\mathbf{f}(\mathbf{p}) = (\mathbf{I} + \sin\varphi\,\mathbf{e}_2 \wedge \mathbf{e}_1 - (1-\cos\varphi)\,\check{\mathbf{I}})\mathbf{p}, \quad \varphi = \zeta\,\tau, \tag{3}$$

where $\mathbf{e}_2 \wedge \mathbf{e}_1 = \mathbf{e}_2 \otimes \mathbf{e}_1 - \mathbf{e}_1 \otimes \mathbf{e}_2$, φ is the *torsion angle*, and τ is the unit torsion angle. From now on, we parametrize the torsional deformation (3) through the torsion angle $\hat{\varphi} = h\,\tau$ of the basis $\mathscr{D} \times \{h\}$. The geometrical structure of the cylinder-like body $\mathscr{C}$ is retained under the deformation (3) and, denoted with $\mathbf{x} \in \mathscr{V}$ the place of x with respect to o, it holds:

$$\mathbf{x} = \mathbf{x}_o + \zeta\,\mathbf{e}_3, \quad \mathbf{x}_o = r(\cos\varphi\,\mathbf{n}(\theta) + \sin\varphi\,\mathbf{n}_{,\theta}(\theta)). \tag{4}$$

We note that, $(\mathbf{x}_o \cdot \mathbf{x}_o)^{1/2} = r$, that is, as expected, the deformation (3) preserves the radius of the cylinder-like body. The deformation gradient $\mathbf{F}$ corresponding to (3) may be represented as

$$\mathbf{F} = \check{\mathbf{F}} + \tau\mathbf{x}_o^* \otimes \mathbf{e}_3 + \mathbf{e}_3 \otimes \mathbf{e}_3, \tag{5}$$

with $\check{\mathbf{F}} \in \mathbb{L}\mathrm{in}(\mathscr{U})$ and $\mathbf{x}_o^* \in \mathscr{U}$ given by

$$\check{\mathbf{F}} = \cos\varphi\,\check{\mathbf{I}} + \sin\varphi\,(\mathbf{e}_2 \wedge \mathbf{e}_1), \quad \mathbf{x}_o^* = \mathbf{e}_3 \times \mathbf{x}_o. \tag{6}$$

From equations (5) and (6) it is easy to verify that the deformation (3) identically satisfies the incompressibility condition $J = \det\mathbf{F} = 1$. We describe the material response of the body through the strain energy function

$$\psi = c_1(I_1 - 3) + c_2(I_2 - 3) + \frac{1}{2}c_1\gamma(I_4 - 1)^2, \quad J = 1, \tag{7}$$

with I_1 and I_2 the linear and quadratic invariant of the Cauchy-Green tensor $\mathbf{C} = \mathbf{F}^T\mathbf{F}$ corresponding to $\mathbf{F}$ and $I_4 = \mathbf{Ce} \cdot \mathbf{e}$. The strain energy (7) describes the mechanical response of an isotropic and incompressible base material with uniaxial reinforcement (see [3]). The corresponding Cauchy stress is

$$\mathbf{T} = 2(\psi_1 + I_1\,\psi_2)\mathbf{B} - 2\psi_2\mathbf{B}^2 + 2\bar{\psi}_4(I_4 - 1)\mathbf{Fe} \otimes \mathbf{Fe} - p\mathbf{I}, \tag{8}$$

with $\psi_\alpha = \frac{\partial\psi}{\partial I_\alpha} = c_\alpha$, $(\alpha = 1,2)$, $\mathbf{B} = \mathbf{FF}^T$, and $\bar{\psi}_4 = c_1\gamma$; the Cauchy stress corresponding to an isotropic and incompressible material turns out by setting $\bar{\psi}_4 = 0$ (corresponding to $\gamma = 0$). We have

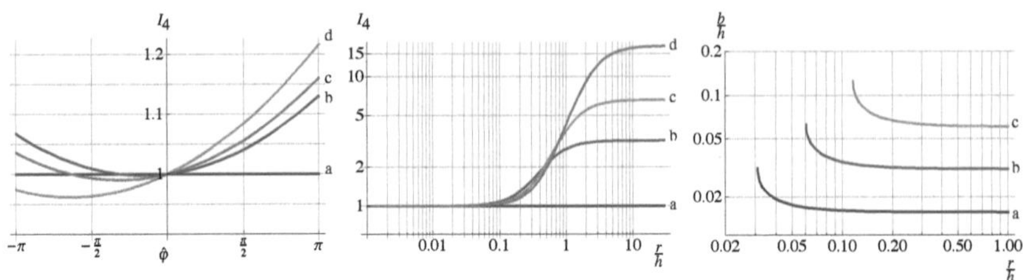

Fig. 1 I_4 VS $\hat{\varphi}$ at r/h=0.1 (left) and I_4 VS r/h at $\hat{\varphi} = \pi/2$ (centre, log-log plot) for four different pitch/height ratios: b/h=0 (a), 1/2 (b), 1 (c), 2 (d). b/h VS r/h (right, log-log plot) to achieve a constant fiber strain along the radius in a torsion with $\hat{\varphi} = \pi/2$: $I_4 = 1.05$ (a), 1.10 (b), 1.20 (c)

$$I_1 = 3 + \tau^2 r^2, \qquad I_4 = 1 + \frac{r^2}{b^2 + r^2}(2 + b\tau)b\tau. \tag{9}$$

The behavior of I_4 is discussed in figure 2. As first, we note on the left that for circumferential fibers (b/h=0) $I_4 \equiv 1$, that is, the fibers length does not change under the torsional deformation (3); otherwise, in a counterclockwise torsional rotation the stretch of the fibers is monotone increasing with b/h. The log-log plot in the center of Fig. 2 shows the dependence of I_4 on r/h for a given torsional angle $\hat{\varphi} = \pi/2$: for the pitch/height ratios considered, the fibers' stretch is quite low for $r << h$ and, interestingly, has a large plateau for $r >> h$. It is worth noting the possibility of winding the fibers around the cylinder with a variable pitch so that, for a given torsion, the stretch remains constant along the radius; right plot in Fig. 2 shows pitch rate b/h versus r/h that realize three different constant stretches in the fiber for $\hat{\varphi} = \pi/2$.

The Balance Equations and the Stress Field

The peculiar geometric structure of the body suggests to using the additive decomposition introduced in the representation formula of the deformation gradient **F** and writing the Cauchy stress as

$$\mathbf{T} = \check{\mathbf{T}} + 2\mathrm{sym}\,(\mathbf{t} \otimes \mathbf{e}_3) + \sigma \mathbf{e}_3 \otimes \mathbf{e}_3, \tag{10}$$

in terms of the plane stress component $\check{\mathbf{T}} \in \mathbb{L}\mathrm{in}(\mathscr{U})$, the plane tangential stress vector $\mathbf{t} \in \mathscr{U}$, and the axial stress component $\sigma \in \mathscr{R}$. The standard balance equations of mechanics may be decomposed in agreement with the geometrical structure of $\mathscr{C}$ into a plane vector component and an axial scalar component as

$$\mathrm{div}\check{\mathbf{T}} + \mathbf{t}' = \mathbf{0} \quad \text{and} \quad \sigma' + \mathrm{div}\mathbf{t} = 0, \tag{11}$$

respectively, where div is the divergence operator in $\mathscr{U}$ and a prime denotes the derivative with respect to ζ. When **T** is given by the equation (8), we find:

$$\check{\mathbf{T}} = (-p + 2(\psi_1 + \psi_2(I_1 - 1)))\check{\mathbf{I}}$$
$$+ (2\tau^2\psi_1 + 2\bar{\psi}_4(I_4 - 1)\frac{(1+b\tau)^2}{(b^2 + r^2)})\mathbf{x}_o^* \otimes \mathbf{x}_o^*, \qquad (12)$$

$$\mathbf{t} = 2((\psi_1 + \psi_2)\tau + 2\bar{\psi}_4(I_4 - 1)b\frac{(1+b\tau)}{(b^2 + r^2)})\mathbf{x}_o^*, \qquad (13)$$

$$\sigma = -p + 2(\psi_1 + 2\psi_2) + 2\bar{\psi}_4(I_4 - 1)\frac{b^2}{(b^2 + r^2)}. \qquad (14)$$

So, it is easy to check from equation (13) that $\mathrm{div}\,\mathbf{t} = 0$ and $\mathbf{t}' = \mathbf{0}$; moreover, equation (14) shows that $\sigma' = p'$ and equations (11) yield

$$\mathrm{grad}\,p = -(2(\psi_1 - 2\psi_2)\tau^2 + 2\bar{\psi}_4 b\tau(1+b\tau)^2(2+b\tau)\frac{r^2}{(b^2 + r^2)^2})\mathbf{x}_o, \quad (15)$$

$$-p' = 0. \qquad (16)$$

With this, we have that $p = \hat{p}(r)$ and

$$\hat{p}(r) = p_o - (\psi_1 - 2\psi_2)\tau^2 r^2$$
$$- \bar{\psi}_4 b\tau(1+b\tau)^2(2+b\tau)(\frac{b^2}{b^2 + r^2} + \log(b^2 + r^2)), \qquad (17)$$

with p_o determined by the boundary conditions.

One of the goals of the work of R.S. Rivlin ([7]) was to find which kind of tractions, if any, should be applied on the boundary of the cylinder-like region $\mathscr{C}$ which were consistent with the torsional deformation (3). He found that a no-traction condition on $\partial\mathscr{D}_o \times \mathscr{I}$ is consistent with a pure torsional deformation; but, a pressure field $P_i\mathbf{m}$ has to be applied on $\partial\Lambda_o \times \mathscr{I}$ if the lacuna Λ_o is present. For a helically

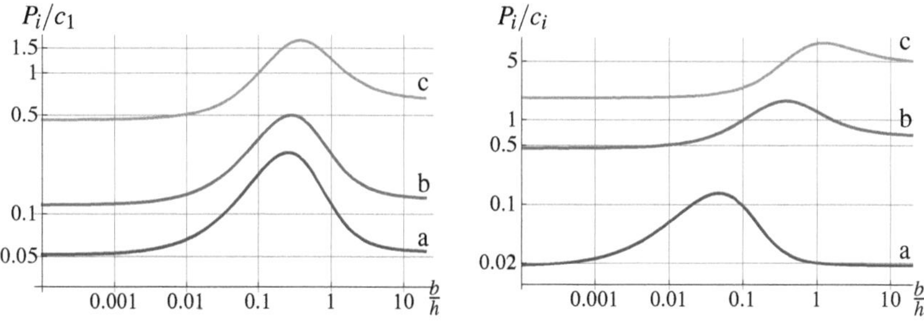

Fig. 2 Inflating pressure VS b/h for three different torsion angles: $\hat{\phi} = \pi/6$ (a), $\pi/4$ (b), $\pi/2$ (c), with $R_o/h = 0.5$, $R_i/R_o = 0.5$, $\gamma = 1$ (left). Inflating pressure VS b/h for three different aspect ratios of the pipe: $R_o/h = 0.1$ (a), 0.5 (b), 1 (c), with $R_i/R_o = 0.5$, $\gamma = 1$, $\hat{\phi} = \pi/2$ (right)

wound fiber-reinforced cylinder-like region, we find analogous results. In absence of the lacuna Λ_o, the no-traction condition $\mathbf{Tm} = \mathbf{0}$ on $\partial \mathcal{D}_o \times \mathcal{I}$ may be written as $\check{\mathbf{T}}\mathbf{m} = \mathbf{0}$ on $\partial \mathcal{D}_o \times \mathcal{I}$; then, equation (17) turns out

$$p_o = 2(\psi_1 + 2\psi_2) + \tau^2 R_o^2 \psi_1$$

$$+ \ \bar{\psi}_4 \, b\tau(2 + b\tau)(1 + b\tau)^2 \left(\frac{b^2}{b^2 + R_o^2} + \log(b^2 + R_o^2)\right); \tag{18}$$

and, with this, the Cauchy stress is completely determined. In presence of the lacuna Λ_o, the no-traction condition on $\partial \mathcal{D}_o \times \mathcal{I}$ is still consistent with the pure torsional deformation (3) and determines completely the Cauchy stress. Thus, it happens that on the boundary $\partial \Lambda_o \times \mathcal{I}$ there is a traction field $\mathbf{Tm} = -P_i\mathbf{m}$ different from zero with

$$P_i = \psi_1 (R_o^2 - R_i^2)\left(\frac{\hat{\varphi}}{h}\right)^2$$

$$+ \ \bar{\psi}_4 \left(b^2 \frac{R_i^2 - R_o^2}{(b^2 + R_i^2)(b^2 + R_o^2)} + \log\frac{b^2 + R_o^2}{b^2 + R_i^2}\right)\frac{b}{h}\hat{\varphi}\left(1 + \frac{b}{h}\hat{\varphi}\right)^2\left(2 + \frac{b}{h}\hat{\varphi}\right). \tag{19}$$

It means that a pure torsional deformation can be maintained on a cylindrical pipe if on the inner mantle a pressure constant field $P_i = \hat{P}_i(R_i, R_o, b, \tau, c_1, \gamma)$ is applied. The pressure field P_i is a fourth order polynomial function of $\hat{\varphi}$, is a non monotone function of the pitch ratio b/h, and contains a stiffness term depending on the geometric characteristics of the pipe R_o/h, R_i/h. Figure 2 summarizes the interesting features of P_i; the non dimensional ratio P_i/c_1 versus b/h is represented in a log-log plot for three different torsion angles (left) and for three different aspect ratios of the pipe (right). On the bases of the cylindrical structure, we find a traction field $\mathbf{T}_\pm \mathbf{e}_3 = \sigma_\pm \mathbf{e}_3 + \mathbf{t}_\pm$ with $\sigma_\pm$ and $\mathbf{t}_\pm$ denoting the values attained by the corresponding fields σ, and $\mathbf{t}$ on the bases $\mathcal{D} \times \{h\}$ and $\mathcal{D} \times \{0\}$, respectively (and $\mathbf{T}_\pm$ is the corresponding Cauchy stress field on the bases)[2]. The Rivlin's result may be recovered by setting $\bar{\psi}_4 = 0$ in the equations (17) and (19). Moreover, we note that for $b = 0$ (circumferential fibers) the stress $\mathbf{T}$ due to a pure torsional deformation does not depend on $\bar{\psi}_4$. It means that circumferential fibers does not reinforce the cylindrical structure with respect to a torsional deformation.

3 Stress Resultants and Torsion Tests

We measure both the torque due to the tractions $\mathbf{t}_\pm$ inducing the torsional deformation and the axial force due to the restrained axial displacement of the structure. The aim is the evaluation of the dependence of the torque $\mathbf{M} = M\mathbf{e}_3$ and the axial force $\mathbf{N} = N\mathbf{e}_3$ on the torsion angle $\hat{\varphi}$, on the modulus γ and on the pitch b of the helix

---

[2] Their role is well different: the axial stress $\sigma_\pm$ are reactive stresses corresponding to the kinematical constraint on the axial displacement; the tangential stress $\mathbf{t}_\pm$ represent the stress fields to be applied on the bases with the aim to produce a pure torsional deformation.

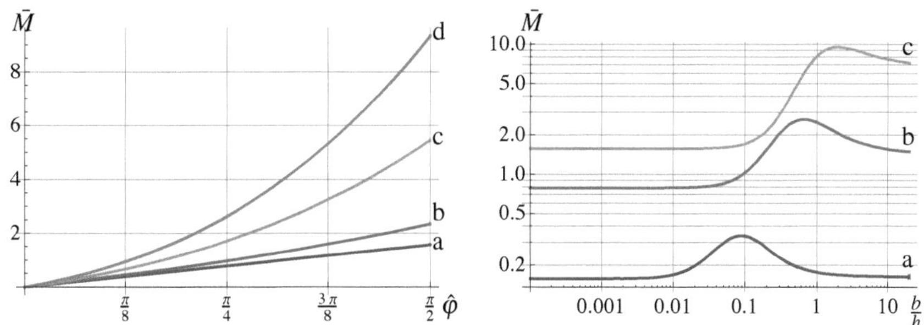

Fig. 3 Torque VS $\hat{\varphi}$ for different fiber stiffness: $\gamma = 0$ (a), 0.1 (b), 0.5 (c), 1 (b) (left); Torque VS b/h for different aspect ratios of the cylinder: Ro/h=0.1 (a), 0.5 (b), 1 (c) (right)

for different cylinder-like structures. The torque and the axial force are determined by the scalar fields

$$M = \int_{\mathscr{P}} \mathbf{x}_o \times \mathbf{t}\, d\mathbf{x}_o \cdot \mathbf{e}_3, \quad \text{and} \quad N = \int_{\mathscr{P}} \sigma\, d\mathbf{x}_o. \tag{20}$$

By using equations $(4)_2$ and (13), we find

$$M = \pi R_o^3 c_1 \left(\frac{R_o}{h}\, \hat{\varphi} + 8\,\gamma\, \frac{R_o}{h}\, f(R_o, b)\, \hat{\varphi} \left(1 + \frac{b}{h}\, \hat{\varphi}\right) \left(2 + \frac{b}{h}\, \hat{\varphi}\right) \right), \tag{21}$$

with

$$f(R_o, b) = \frac{b^2}{R_o^2} \log \frac{b^2}{b^2 + R_o^2} + \frac{R_o^2 + 2b^2}{2(b^2 + R_o^2)}. \tag{22}$$

In (21), the first addendum is the torque competing to an isotropic and incompressible cylinder, a linear function of $\hat{\varphi}$, corresponding to the Rivlin solution; the second addendum defines the correction due to the presence of the helicoidal fibers. The

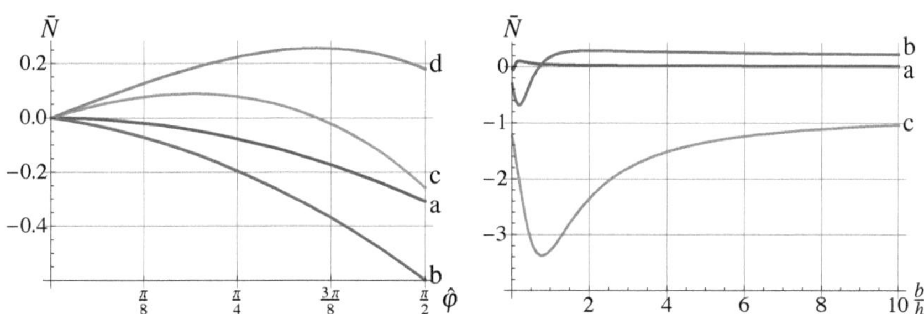

Fig. 4 Cylinder. Axial force VS $\hat{\varphi}$ for different pitch/height ratios: b/h=0 (a), 0.1 (b), 0.5 (c), 1 (d) (left). Axial force VS pitch/height at $\hat{\varphi} = \pi/2$ for different aspect ratios of the cylinder: $R_o/h = 0.1$ (a), 0.5 (b), 1 (c) (right). $\gamma = 1$ in both plots

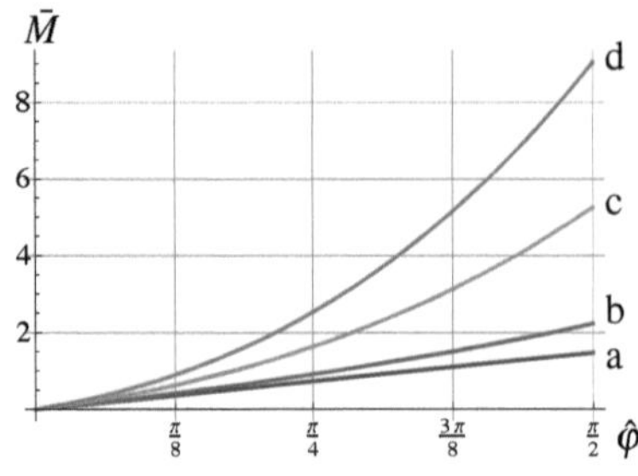 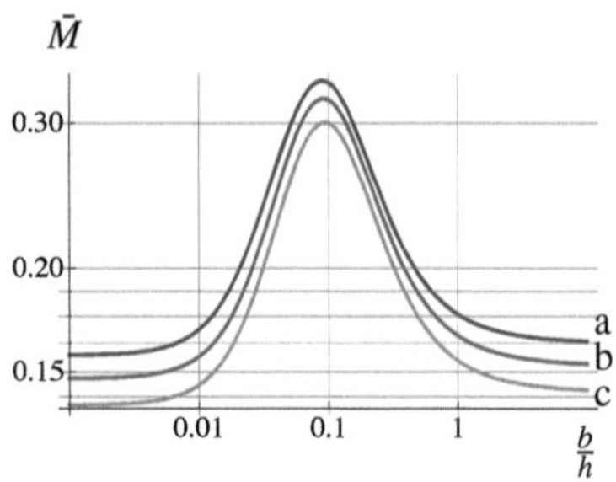

Fig. 5 Pipe. Torque VS $\hat{\varphi}$ for different fiber stiffness $\gamma = 0$ (a), 0.1 (b), 0.5 (c), 0.6 (d). Right: Torque VS pitch/height for different wall thickness of the tube: Ri/Ro=0.1 (a), 0.5 (b), 0.6 (c); Ro/h=1, $\gamma = 1$, $\varphi_h = \pi/2$ (right)

dependence of the dimensionless torque $\bar{M} = M/\pi R_o^3 c_1$ on the relevant parameters is shown in figure 3: $\bar{M}(\hat{\varphi})$ is an increasing monotone function with respect to $\hat{\varphi}$ and γ (left); more interesting, for any given $\hat{\varphi}$, $\bar{M}$ is not monotone with b/h, and the maximum depends on R_o/h, a parameter measuring the slenderness of the cylinder (right). As far as the axial force N is concerned, we find

$$N = -\frac{1}{2}\pi R_o^2 c_1 \left(\frac{R_o}{h}\right)^2 \hat{\varphi}^2$$
$$+ 2\pi R_o^2 c_1 \gamma \hat{\varphi} \left(2 + \frac{b}{h}\hat{\varphi}\right)\left(g_1(R_o,b) - \frac{1}{2}\left(1 + \frac{b}{h}\hat{\varphi}\right)^2 f_1(R_o,b)\right) \tag{23}$$

with

$$f_1(R_o,b) = 2f(R_o,b), \quad g_1(R_o,b) = \frac{R_o^2}{2(b^2 + R_o^2)} - f(R_o,b).$$

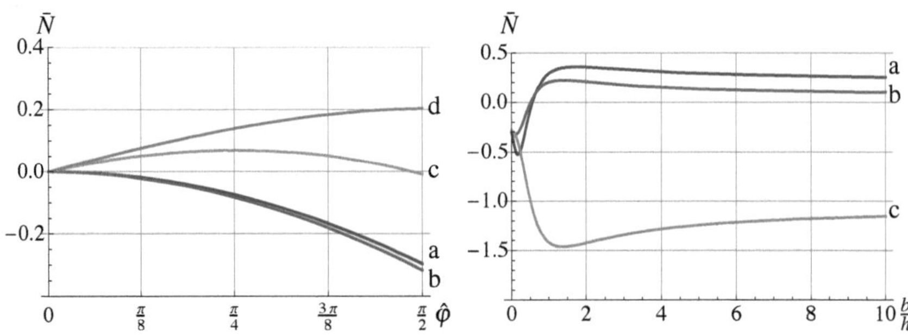

Fig. 6 Pipe. Axial force VS $\hat{\varphi}$ for different pitch/height ratios: b/h=0 (a), 0.1 (b), 0.5 (c), 1 (d), with $R_o/h = 0.5$, $R_i/R_o = 0.2$ (left). Torque VS b/h for different wall thickness of the tube: $R_i/R_o = 0.1$ (a), 0.2 (b), 0.3 (c), with $R_o/h = 0.5$, $\gamma = 1$, $\varphi_h = \pi/2$ (right). $\gamma = 1$ in both plots

The first row of equation (23) defines the axial force competing to the isotropic and incompressible cylinder; as it is expected, in the linear approximation (that is, for small $\hat{\varphi}$) this term goes to zero. The second row gives the contribution to the axial force due to the anisotropy; it is worth noting that such addendum is different from zero in a linear theory. Figure 4 shows the dependence of the dimensionless axial force $\bar{N} = N/\pi R_o^2 c_1$ on $\hat{\varphi}$ for different b/h, and for cylinders with $R_o/h = 0.5$ (left), and $\bar{N}$ on b/h for different slenderness parameters, at $\hat{\varphi} = \pi/2$ (right).

The same analysis is performed for a pipe; in such a case, while $\bar{M}(\hat{\varphi})$ is always an increasing monotone function with respect to $\hat{\varphi}$ and γ, Fig. (5, left), more interesting in a pipe is the behaviour of $\bar{M}$ with respect to b/h at a fixed $\hat{\varphi}$ and γ ((5), right); as figure (5) evidences, it exists a specific and small range of b/h where $\bar{M}$ attains large values, actually independent on the ratio R_i/R_o. Dependence of the dimensionless axial force $\bar{N}$ on $\hat{\varphi}$ or on b/h is very similar for cylinders and pipes, cfr. Fig. (4) and (6).

References

1. de Botton, G., Hariton, I., Socolsky, E.A.: Neo-Hookean fiber-reinforced composites in finite elasticity. J. Mech. Phys. Solids 54, 533–559 (2006)
2. Merodio, J., Saccomandi, G., Sgura, I.: The rectilinear shear of fiber–reinforced incompressible non-linearly elastic solids. Int. J. Non-Lin. Mech. 42, 342–354 (2007)
3. Merodio, J., Ogden, R.W.: Mechanical response of fiber-reinforced incompressible non-linearly elastic solids. International Journal of Non-Linear Mechanics 40, 213–227 (2005)
4. Merodio, J., Ogden, R.W.: Tensile instabilities and ellipticity in fiber-reinforced compressible non-linearly elastic solids. International Journal of Engineering Science 43, 697–706 (2005)
5. Merodio, J., Ogden, R.W.: On tensile instabilities and ellipticity loss in fiber-reinforced incompressible non-linearly elastic solids. Mechanics Research Communications 32, 290–299 (2005)
6. Ogden, R.W.: Nonlinear elasticity, anisotropy, material stability and residual stresses in soft tissue. In: Holzapfel, G.A., Ogden, R.W. (eds.). CISM Courses and Lectures Series, vol. 441, pp. 65–108. Springer, Wien (2003)
7. Rivlin, R.S.: A note on the torsion of an incompressible highly-elastic cylinder. Proc. Cambridge Philos. Soc. 45, 485–487 (1949)
8. Spencer, A.J.M.: Continuum Theory of the Mechanics of Fibre-Reinforced Composites. Springer, Heidelberg (1984)

Essentially Nonlinear Strain Waves in Solids with Complex Internal Structure

Alexey V. Porubov

Abstract. Analytical modeling of essentially nonlinear strain waves in solids is developed based on a proper assumption about their complex internal structure. The exact travelling solitary wave solutions are used as a tool to study nonlinear and dispersive features of the solid. Various scenarios are found of the propagation of the macro-strain localized waves and corresponding waves of structural deviations. It is found by explicit analytical relationships that they are realized depending upon the velocity of the wave. This may help to study the material properties by dynamical methods.

Introduction

Many phenomena cannot be explained without taking into account an internal structure of a material. It might be micro- or nano-structures in metals, alloys, soft tissues etc. as well as macro- structures of the rocks and soils. These complex internal structures are modeled both by pure phenomenological approaches and the models based on proper assumptions about a behavior of an internal structure, e.g., translational or rotational motion of internal elements. Application of the models requires knowledge of the values of their parameters. Most of modern models are non-linear, and the parameters characterizing nonlinear features often cannot me measured using static methods. Application of the dynamic methods, e.g. acoustical, is not easy in the non-linear case. There is a prominent class of non-linear strain waves that may propagate keeping their shape and velocity. One of them is the bell-shaped solitary wave arising as a result of a balance between nonlinearity and dispersion. It is possible to obtain the relationships between the parameters of

Alexey V. Porubov
Institute of Problems in Mechanical Engineering, Bolshoy av., 61, V.O.,
Saint-Petersburg 199178, Russia
e-mail: porubov.math@mail.ioffe.ru

J.-F. Ganghoffer, F. Pastrone (Eds.): Mech. of Microstru. Solids, LNACM 46, pp. 119–126.
springerlink.com

the wave and the parameters of the material in an explicit form. On the one hand, it allows us to find the conditions required for the existence of the strain solitary wave. On the other hand, observation of such localized waves allows us to estimate the unknown material parameters using the measurements of the wave amplitude and velocity.

So, the first problem is to develop suitable non-linear model with dispersion. The classic elastic theory admits two main sources of nonlinearity. The first one is the geometrical nonlinearity described by the exact relationship for the tensor of finite strains. Another one, the physical nonlinearity, is caused by an anharmonicity in an interatomic interaction. Contrary to the geometrical nonlinearity, it is not described by an exact analytical formula but is modeled using some hypothesis about deformations. Of most popular are the weakly non-linear models based on the power series expansion of the energy density in small strains or strain tensor invariants. The Murnaghan model [1] may be noted among them since it is valid for isotropic materials like metals, polymers etc. In particular, the series truncated by the fourth order term is called the nine-constants Murnaghan model, and the energy density Π reads

$$\Pi = \frac{\lambda + 2\mu}{2} \; I_1^2 - 2\mu \; I_2 + \frac{l + 2m}{3} \; I_1^3 - 2m \; I_1 I_2 + n \; I_3 + $$
$$\nu_1 \; I_1^4 + \nu_2 \; I_1^2 I_2 + \nu_3 \; I_1 I_3 + \nu_4 \; I_2^2, \tag{1}$$

where $I_k, k = 1, 2, 3$ - are the invariants of the Cauchy-Green strain tensor, the fourth order elastic moduli (ν_1, ν_2, ν_3, ν_4), as well as the third order ones, l, m, n, may be of either sign contrary to the positive second order moduli λ μ. One obtains the stress-strain relationship from Eq.(1) in the 1D case, $P = EU_x + C_1 U_x^2 + C_2 U_x^3$, ($U_x$ is longitudinal strain, E is the Young modulus), where the second term describes the so-called quadratic nonlinearity, and $C_1 = C_1(l, m , n)$. The last term accounts for the so-called cubic nonlinearity, and its coefficient C_2 depends on both the third and the fourth order moduli. Usually, only quadratic nonlinearity is used (the so-called five-constants Murnaghan model, $\nu_i = 0$ in Eq.(1)) to describe longitudinal strain waves since relative contribution, $C_2 U_x / C_1$, of the last term in the stress-strain relationship is negligibly small for usual elastic materials [2, 3, 4, 5, 6].

Dispersion in classic elastic bodies is caused by finite transverse sizes of a waveguide. To see how strong this dispersion may be, one can note experimental observations of the strain solitary wave of the amplitude of order 10^{-4} and the width $33mm$ in a rod of the radius $R = 5mm$ [5]. Also dispersion appears as a result of an internal complex structure of material [7, 8, 9, 10]. Some estimations have been done for materials with grains and for sandstones [11, 12]. According to them, the solitary waves may exist in such media with anticipated typical width $0.1 - 100m$. Finally, one can note phonon dispersion arising in crystals due to the finite atomic sizes. It gives rise to the strain solitary wave of the amplitude of order 10^{-4} and the width of $100\overset{\circ}{A}$ observed in experiments [13]. One can note that all above mentioned experiments were described by the solitary wave solutions obtained using Eq.(1) [5, 13].

Recently a phenomenon of abnormal nonlinearity has been discovered for the rocks, caused by the presence of the components with contrasting elastic features including cracks, intergranular contacts, dislocations at the boundaries of the grains of poly-crystals [14, 15, 16]. The power series truncation like (1) cannot describe this essentially nonlinear process. However, its *formal* use in Refs. [14, 15, 16, 17] is applicable since it allowed them to define the values of the third and the fourth order moduli for some rocks and soils. The most important is that it turns out that the contribution of the quadratic and cubic nonlinearities is of the same order for seismic materials. One can call the formal use of the expressions from weakly nonlinear theory the phenomenological approach. Another approach takes into account proper internal structures. Thus, the rotatory molecular groups were added to the usual one atomic chain in Refs. [3, 18], and large rotations were considered. A more complicated internal motion is modeled in [19, 20], where translational internal motion is taken into account together with rotations. The main attention was focused mainly on the micro-motions while macro-strain solitary waves were not considered in detail in Refs. [3, 18, 19, 20].

In this paper, the model developed [19, 20] will be studied using macro-strain solitary wave solutions like it was done before for the weakly nonlinear models [6]. In other words, the attention is paid on how macro-strain wave may affect variations in the internal structure. In Sect.1 it will be shown that propagation of nonlinear longitudinal strain waves is governed by the ODE reduction of the Gardner equation that includes both quadratic and cubic nonlinear terms together with dispersion. Next Section is devoted to an analysis of its solitary wave solutions. It is found that propagation of the solitary wave strongly depends upon its velocity, and similar macro-strain waves give rise to completely different deviations in the internal structure. All findings are summarized in conclusions.

1 Media with Essentially Nonlinear Translational Internal Structure

The model in [19, 20] is based on a complex lattice consisting of two sub-lattices, whose linear analog has been developed by Born & Kun [21] as an extension of the well known Born-Karman model [3]. Besides interatomic forces between atoms, the relative sub-lattices motion is taken into account in the model, hence it generalizes the Frenkel-Kontorova model for the simple lattice to describe structural deviations in the bi-atomic lattice.

According to [19, 20] a continuum approach is developed directly without making a continuum limit of a discrete model like it is done in Refs. [3, 18]. The equations are derived for the vectors of macro-displacement U and relative micro-displacement u for the pair of atoms with masses m_1, m_2,

$$\mathbf{U} \;=\; \frac{m_1\mathbf{U_1} + m_2\mathbf{U_2}}{m_1 + m_2}, \quad \mathbf{u} \;=\; \frac{\mathbf{U_1} - \mathbf{U_2}}{a}$$

where a is a period of sub-lattice. In general, it allows us to describe both translational and rotational motions of the internal structure. In the one-dimensional case, only translational motion is considered, then the kinetic energy density K reads

$$K \;=\; \frac{\rho U_t^2 + \mu u_t^2}{2},$$

and the internal density energy Π reads

$$\Pi \;=\; \frac{E\,U_x^2 + \kappa\,u_t^2}{2} + (p - SU_x)(1 - \cos(u)) \tag{2}$$

Comparing it with the Murnaghan model (1), one can see the absence of the terms describing physical nonlinearity at the macro level. Instead, the last term in Eq. (2) accounts for the sub-lattices interaction. In the absence of coupling at $S \;=\; 0$, the Frenkel-Kontorova model is revealed. Trigonometric functions allow us to describe an identity of the complex lattice after displacement proportional to to the period of the sub-lattice.

The Hamilton principle is employed to obtain the following coupled equations,

$$\rho U_{tt} - E\,U_{xx} \;=\; S(\cos(u) - 1)_x, \tag{3}$$

$$\mu u_{tt} - \kappa u_{xx} \;=\; (SU_x - p)\sin(u), \tag{4}$$

For wave processes the r.h.s. in Eqs. (3), (4) should be of lower order in comparison with those in the l.h.s. The function u may of order one, the macro-strain, $v \;=\; U_x$, is of order $10^{-4} \div 10^{-5}$, and the Young modulus is of order 10^{10} for most of materials. Therefore, the value of the parameter S is of order 10^6 as follows from Eq. (3) while $p \sim 10^3$ follows from Eq. (4).

One can obtain from Eqs. (3), (4) the single equation for the macro-strain $v \;=\; U_x$ considering travelling wave solutions depending only on the phase variable $\theta \;=\; x - V\,t$,

$$v_\theta^2 \;=\; a_1\,v + a_2\,v^2 + a_3\,v^3 + a_4\,v^4. \tag{5}$$

where the coefficients a_i depend on the velocity V and the constant of integration σ, arising after integration of Eq.(3),

$$a_1 \;=\; \frac{2p\,\sigma(2S + \sigma)}{S(E - \rho\,V^2)(\mu\,V^2 - \kappa)}, \quad a_2 \;=\; \frac{p(4p(E - \rho\,V^2)(S + \sigma) + S\,\sigma(2S + \sigma))}{S(E - \rho\,V^2)(\kappa - \mu\,V^2)},$$

$$a_3 \;=\; \frac{2(p(E - \rho\,V^2) + S(S + \sigma))}{S(\mu\,V^2 - \kappa)}, \quad a_4 \;=\; \frac{E - \rho\,V^2}{\kappa - \mu\,V^2}$$

This equation possesses solutions vanishing at infinity, $\mid \theta \mid \to \infty$, provided that $a_1 = 0$ that happens for $\sigma = 0$ or for $\sigma = -2S$. This wave arises as a result of the balance between dispersion, v_θ^2, and quadratic and cubic nonlinearities, all caused by the coupling, $S \neq 0$. The nonlinear term coefficients ratio $a_3/a_4 = \max\{p/S, S/E\}$, turns out of order 10^{-3}, that yields almost equal contribution of the quadratic and cubic nonlinearities for typical elastic strains of order 10^{-4} similar to the case of abnormal nonlinearity in the seismic media [15, 16].

In the weakly nonlinear case, the ODE with quadratic nonlinear term is obtained, and the non-linear term coefficient does not depend on the phase velocity [6].

2 Analysis of Localized Strain Wave Solutions

The equation (5) at $a_1 = 0$ possesses known exact solutions of two types,

$$v_1 = \frac{A}{Q \cosh(k\,\theta) + 1}, \tag{6}$$

$$v_2 = -\frac{A}{Q \cosh(k\,\theta) - 1}. \tag{7}$$

where for $\sigma = 0$

$$A = \frac{4\,S}{\rho(c_0^2 + c_L^2 - V^2)}, Q_\pm = \pm\frac{c_L^2 - V^2 - c_0^2}{c_L^2 - V^2 + c_0^2}, k = 2\sqrt{\frac{p}{\mu(c_l^2 - V^2)}} \tag{8}$$

and for $\sigma = -2S$

$$A = \frac{4\,S}{\rho(c_0^2 + V^2 - c_L^2)}, Q_\pm = \pm\frac{V^2 - c_L^2 - c_0^2}{V^2 - c_L^2 + c_0^2}, k = 2\sqrt{\frac{p}{\mu\,(V^2 - c_l^2)}} \tag{9}$$

where $c_L^2 = E/\rho$, $c_0^2 = S^2/(p\,\rho)$, $c_l^2 = m/\mu$. Hence in the first case $V^2 < c_l^2$, and in the second one - $V^2 > c_l^2$. In both cases the solution of the first kind (6) is realized for $Q_\pm > 0$, while solution of the second kind (7) appears at $Q_\pm < -1$. The sign of the amplitude of the wave $A/(Q_\pm + 1)$ depends on the relation between V, c_L, c_0. The variation of the value of the amplitude of the wave $A/(Q_- + 1)$ does not depend on the variation of velocity V while $A/(Q_+ + 1)$ does.

An analysis of the exact solutions should take into account the coupling governed by Eq. (3). Then it follows that simultaneous existence of the compression and tensile waves is impossible that differs from the case of abnormal nonlinearity studied in [22, 23, 24]. Also different micro deviations are caused by similar bell-shaped profiles of the macro-strain wave. This is illustrated in Figs. 1,2 for the case $\sigma = 0$ corresponding to the initially undisturbed internal structure. One can see in Fig.1 that the compression macro-wave

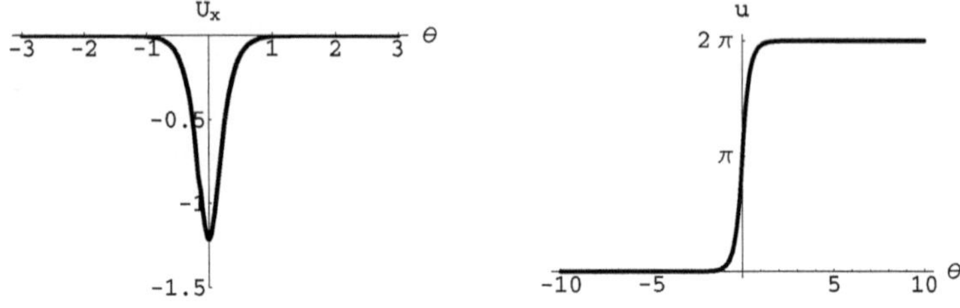

Fig. 1 Compression wave of macro-strain and corresponding micro-strain waves for the velocities from the interval $c_L^2 + c_0^2 < V^2 < c_l^2$

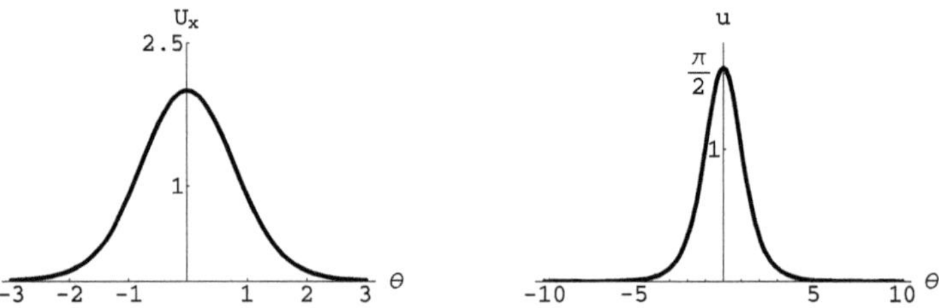

Fig. 2 Compression wave of macro-strain and corresponding micro-strain waves for the velocities from the interval $c_L^2 - c_0^2 < V^2 < c_l^2,\ V^2 < c_l^2$

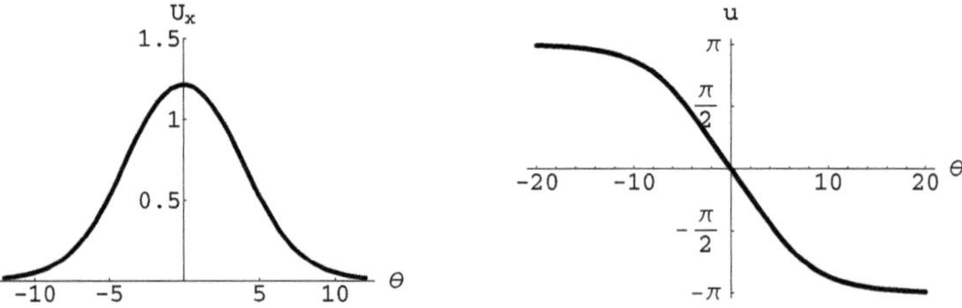

Fig. 3 Tensile wave of macro-strain and corresponding micro-strain waves for the velocities from the interval $c_L^2 + c_0^2 < V^2,\ V^2 > c_l^2$

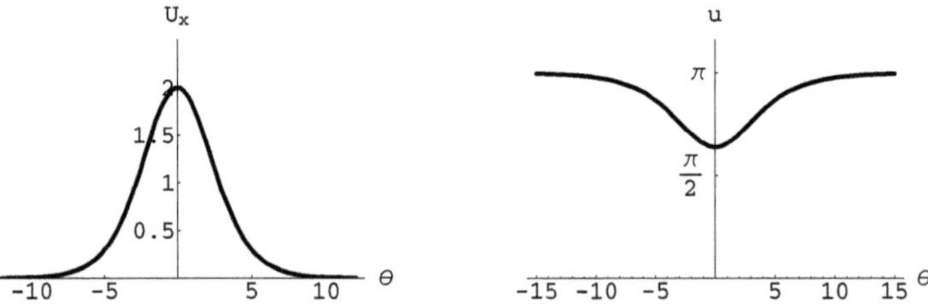

Fig. 4 Tensile wave of macro-strain and corresponding micro-strain waves for the velocities from the interval $c_L^2 < V^2 < c_l^2 + c_0^2,\ V^2 > c_l^2$

gives rise to the transition of the sub-lattice elements by one period. On the contrary, the similar bell-shaped tensile wave in Fig.2 provides perturbation by a quarter of period of the internal structure with coming back to the initial state.

Similar distinctions in the micro-filed profiles are observed for the case $\sigma = -2S$ that corresponds to the internal structure initially disturbed by static loading. This is illustrated in Figs. 3,4. In Fig.3 one disturbed microstructure transforms into another but equivalent one. The similar macro-tensile wave provides only perturbation of the initial disturbed by less than half a period of the sub-lattice.

In both cases, $\sigma = 0, \sigma = -2S$ it is the phase velocity V that defines one or another kind of propagation what is clearly seen in the figure captions.

3 Conclusions

To sum up, the study of the solitary wave behaviour allows us to establish analytically the relationships between the velocity of the wave and the parameters of the material required for the existence of one or another kind of localized strain wave. This may help us to estimate the nonlinear and dispersive properties of the material caused by coupling of the macro- and micro-fields. It is important that the governing equations turns out similar to that of obtained in the framework of the phenomenological approach when weakly nonlinear model (1) is formally used [22, 23, 24]. A similar technique may be employed for the predictions about dynamic localized structures for the model developed in Refs. [3, 18]. In Refs. [22, 23, 24] numerical simulations are performed to demonstrate that the profiles of the exact travelling wave solutions arise even in a more general unsteady processes as a result of an evolution of rather arbitrary initial perturbation. The same results for the model considered will be done in the nearest future.

Acknowledgements. This work has been supported by the grant of the Russian Science Support Foundation.

References

1. Murnaghan, F.D.: Finite Deformations of an Elastic Solid. J. Wiley, New York (1951)
2. Engelbrecht, J.: Nonlinear wave processes of deformation in solids. Pitman, Boston (1983)
3. Maugin, G.A.: Nonlinear Waves in Elastic Crystals. Oxford University Press, Oxford (1999)
4. Ostrovsky, L.A., Potapov, A.I.: Modulated waves in linear and nonlinear media. The John Hopkins University Press, Baltimore (1999)

5. Samsonov, A.M.: Strain soltons in solids and how to construct them. Chapman & Hall/CRC (2001)
6. Porubov, A.V.: Amplification of nonlinear strain waves in solids. World Scientific, Singapore (2003)
7. Erofeev, V.I., Potapov, A.I.: Longitudinal strain waves in non-linearly elastic media with coupled stresses. Int. J. Nonl. Mech. 28, 483–488 (1993)
8. Erofeev, V.I.: Wave processes in solids with microstructure. World Scientific, Singapore (2003)
9. Porubov, A.V., Pastrone, F.: Nonlinear bell-shaped and kink-shaped strain waves in microstructured solids. Intern. J. Non-Linear Mech. 39, 1289–1299 (2004)
10. Engelbrecht, J., Berezovsky, A., Pastrone, F., Braun, M.: Waves in microstructured materials and dispersion. Philos. Mag. 85, 4127–4141 (2005)
11. Savin, G.N., Lukashev, A.A., Lysko, E.M.: Elastic wave propagation in a solid with microstructure. Soviet Appl. Mech. 15, 725–728 (1973)
12. Bykov, V.G.: Nonlinear wave processes in geological media. Dal'nauka, Vladivostok, Russia (2000) (in Russian)
13. Hao, H.-Y., Maris, H.J.: Experiments with acoustic solitons in crystalline solids. Phys. Rev. B. 64, 064302 (2001)
14. Zaitsev, V.Y., Nazarov, V.E., Talanov, V.I.: "Nonclassical" manifestations of microstructure-induced nonlinearities: new prospects for acoustic diagnostics. Physics-Uspekhi 49(1) (2006)
15. Belyaeva, I.Y., Zaitsev, V.Y., Ostrovsky, L.A.: Nonlinear acoustical properties of granular media. Acoustical Physics 39, 11–16 (1993)
16. Belyaeva, I.Y., Zaitsev, V.Y., Ostrovsky, L.A., Sutin, A.M.: Elastic nonlinear parameter as an informative characteristic in problems of prospecting seismology. Izvestia of Academy of Sciences USSR, Physics of Solid Earth 30, 890–894 (1994)
17. Winkler, K.W., Liu, X.: Measurements of third-order elastic constants in rocks. JASA 100, 1392–1398 (1996)
18. Maugin, G.A., Pouget, J., Drouot, R., Collet, B.: Nonlinear electromechanical couplings. John Wiley & Sons, UK (1992)
19. Aero, E.L.: Micromechanics of a double continuum in a model of a medium with variable periodic structure. J. Eng. Math. 55, 81–95 (2002)
20. Aero, E.L., Bulygin, A.N.: Strongly Nonlinear Theory of Nanostructure Formation Owing to Elastic and Nonelastic Strains in Crystalline Solids. Mechanics of Solids 42, 807–822 (2007)
21. Born, M., Kun, H.: Dynamic theory of crystal lattices. Clarendon Press, Oxford (1954)
22. Porubov, A.V., Maugin, G.A.: Longitudinal strain solitary waves in presence of cubic nonlinearity. International Journal of Non-Linear Mechanics 40, 1041–1048 (2005)
23. Porubov, A.V., Maugin, G.A.: Propagation of localized longitudinal strain waves in a plate in presence of cubic nonlinearity. Phys. Rev. E 74, 046617–046624 (2006)
24. Porubov, A.V., Maugin, G.A.: Improved description of longitudinal strain solitary waves. Journal of Sound and Vibration 310/3, 694–701 (2008)

Author Index

MIX
Papier aus verantwortungsvollen Quellen
Paper from responsible sources
FSC® C105338
www.fsc.org

If you have any concerns about our products,
you can contact us on
ProductSafety@springernature.com

In case Publisher is established outside the EU,
the EU authorized representative is:
Springer Nature Customer Service Center GmbH
Europaplatz 3, 69115 Heidelberg, Germany

Printed by Libri Plureos GmbH
in Hamburg, Germany